Bernhard Marks

Marks' First Lessons in Geometry, Objectively Presented

Bernhard Marks

Marks' First Lessons in Geometry, Objectively Presented

ISBN/EAN: 9783337109509

Printed in Europe, USA, Canada, Australia, Japan

Cover: Foto ©Paul-Georg Meister /pixelio.de

More available books at **www.hansebooks.com**

MARKS'

FIRST LESSONS IN GEOMETRY.

IN TWO PARTS.

OBJECTIVELY PRESENTED,

AND DESIGNED FOR

THE USE OF PRIMARY CLASSES IN GRAMMAR SCHOOLS, ACADEMIES, ETC.

BY

BERNHARD MARKS,

PRINCIPAL OF LINCOLN SCHOOL, SAN FRANCISCO.

NEW YORK:

PUBLISHED BY IVISON, PHINNEY, BLAKEMAN, & CO.

PHILADELPHIA: J. B. LIPPINCOTT & CO.

CHICAGO· S. C. GRIGGS & CO.

1869.

PREFACE.

How it ever came to pass that Arithmetic should be taught to the extent attained in the grammar schools of the civilized world, while Geometry is almost wholly excluded from them, is a problem for which the author of this little book has often sought a solution, but with only this result; viz., that Arithmetic, being considered an elementary branch, is included in all systems of elementary instruction; but Geometry, being regarded as a higher branch, is reserved for systems of advanced education, and is, on that account, reached by but very few of the many who need it.

The error here is fundamental. Instead of teaching the *elements of all branches, we teach elementary branches* much too exhaustively.

The elements of Geometry are much easier to learn, and are of more value when learned, than advanced Arithmetic; and, if a boy is to leave school with merely a grammar-school education, he would be better prepared for the active duties of life with a *little* Arithmetic and *some* Geometry, than with *more* Arithmetic and *no* Geometry.

Thousands of boys are allowed to leave school at the age of fourteen or sixteen years, and are sent into the carpenter-shop, the machine-shop, the mill-wright's, or the surveyor's office, stuffed to repletion with Interest and Discount, but so utterly ignorant of the merest elements of Geometry, that they could not find the centre of a circle already described, if their lives depended upon it.

Unthinking persons frequently assert that young children are incapable of reasoning, and that the truths of Geometry are too abstract in their nature to be apprehended by them.

To these objections, it may be answered, that any ordinary child, five years of age, can deduce the conclusion of a syllogism if it understands the terms contained in the propositions; and that nothing can be more palpable to the mind of a child than forms, magnitudes, and directions.

There are many teachers who imagine that the perceptive faculties of children should be cultivated *exclusively* in early youth, and that the reason should be addressed only at a later period.

It is certainly true that perception should receive a larger share of attention than the other faculties during the first school years; but it is equally certain that *no* faculty can be safely disregarded, even for a time. The root does not attain maturity before the stem appears; neither does the stem attain its growth before its branches come forth to give birth in turn to leaves; but root, stem, and leaves are found simultaneously in the youngest plant.

That the reason may be profitably addressed through the medium of Geometry at as early an age as seven years is asserted by no less an authority than President Hill of Harvard College, who says, in the preface to his admirable little Geometry, that a child seven years old may be taught Geometry more easily than one of fifteen.

The author holds that this science should be taught in all

primary and grammar schools, for the same reasons that apply to all other branches. One of these reasons will be stated here, because it is not sufficiently recognized even by teachers. It is this : —

The prime object of school instruction is to place in the hands of the pupil the means of continuing his studies without aid after he leaves school. The man who is not a student of some part of God's works cannot be said to live a rational life. It is the proper business of the school to do for each branch of science exactly what *is* done for reading.

Children are taught to read, not for the sake of what is contained in their readers, but that they may be able to read all through life, and thereby fulfil one of the requirements of civilized society. So, enough of each branch of science should be taught to enable the pupil to pursue it after leaving school.

If this view is correct, it is wrong to allow a pupil to reach the age of fourteen years without knowing even the alphabet of Geometry. He should be taught at least how to *read* it.

It certainly does seem probable, that if the youth who now leave school with so much Arithmetic, and no Geometry, were taught the first rudiments of the science, thousands of them would be led to the study of the higher mathematics in their mature years, by reason of those attractions of Geometry which Arithmetic does not possess.

TO THE PROFESSIONAL READER.

THIS little book is constructed for the purpose of instructing large classes, and with reference to being used also by teachers who have themselves no knowledge of Geometry.

The first statement will account for the many, and perhaps seemingly needless, repetitions ; and the second, for the *suggestive* style of some of the questions in the lessons which *develop* the matter contained in the review-lessons.

Attention is respectfully directed to the following points : —

First the particular, then the general. See page 25.

Why is *m n g* an acute angle?

What is an acute angle?

Here the attention is directed first to this particular angle ; then this is taken as an example of its kind, and the idea generalized by describing the class. See also page 29.

Why are the lines *ef* and *gh* said to be parallel?

When are lines said to be parallel?

Many of the questions are intended to test the vividness of the pupil's conception. See page 29.

Also page 78. If the circumference were divided into 360 equal parts, would each arc be large or small?

Many of the questions are intended to test the attention of the pupil.

The thing is not to be recognized by the definition ; but the definition is to be a description of the thing, a description of the conception brought to the mind of the pupil by means of the name.

CONTENTS.

PART I.

8   CONTENTS.

PART II.

AXIOMS AND THEOREMS.

FIRST LESSONS IN GEOMETRY.

PART FIRST.

LESSON FIRST.

LINES.

Note to the Teacher. — In all the development-lessons, the pupils are to be occupied with the diagrams, and not with the printed matter.

See Note A, Appendix.

Refer to Diagram 1, and show that
What are here drawn are intended to represent *length* only.
They have a little width, that they may be seen.
They are called *lines*.

A line is that which has length only.

POINTS

Show that
Position is denoted by a point.
It occupies no space.
It has *some* size, that it may be seen.
The ends of a line are points.
A line may be regarded as a succession of points.
The intersection of two lines is a point.
A point is named by placing a letter near it.

2   9

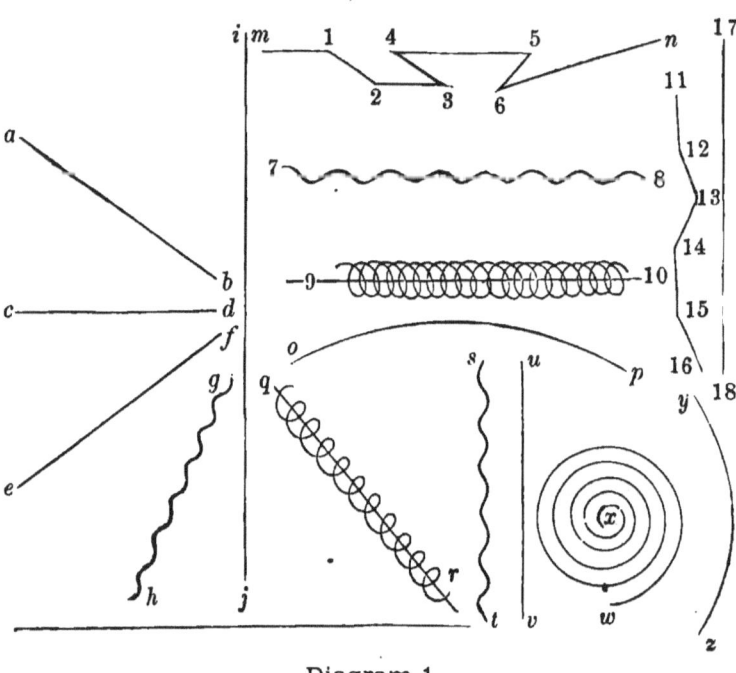

Diagram 1.

A point may be represented by a dot. The point is in the center of the dot.

A point is that which denotes position only.

A line is named by naming the points at its ends. Read all the lines in Diagram 1.

CROOKED LINES.

See Note B, Appendix.

Does the line *m n* change direction at the point 1 ? At what other points does it change direction? It is called a crooked line.

A crooked line is one that changes direction at some of its points.

CURVED LINES.

The line *o p* changes direction at every point.
It is called a curved line.

A curved line is one that changes direction at every
point.

STRAIGHT LINES.

Does the line *i j* change direction at any point?
It is called a straight line.

A straight line is one that does not change direction
at any point.

OTHER LINES.

The line *q r* winds about a line.
It is called a *spiral line*.
The line *w x* winds about a point.
It also is called a spiral line.

A spiral line is one that winds about a line or point.

The line 7 8* looks like waves.
It is called a wave line.

What kind of a line is *a b?*
Why?   What is a straight line?
What kind of a line is 11 16?
Why?   What is a crooked line?
What kind of a line is *o p?*
Why?   What is a curved line?

* To be read seven, eight, not seventy-eight.

What kind of a line is *s t ?*
Why ?
What kind o* a line is 9 10 ?
Why ?   What is a spiral line ?
What kind of a line is *w x ?*
Why ?

———◆———

LESSON SECOND.

REVIEW.

Read all the straight lines.   (DIAGRAM 2.)
Why is *m n* a straight line ?
Define a straight line.
Read all the crooked lines.
Why is 7 8 a crooked line ?
Define a crooked line.
Read all the curved lines.
Why is 5 6 a curved line ?
What is a curved line ?
Read all the wave lines.
Read all the spiral lines.
Why is 3 4 a spiral line ?
Why is *u v* a spiral line ?
What is a spiral line ?

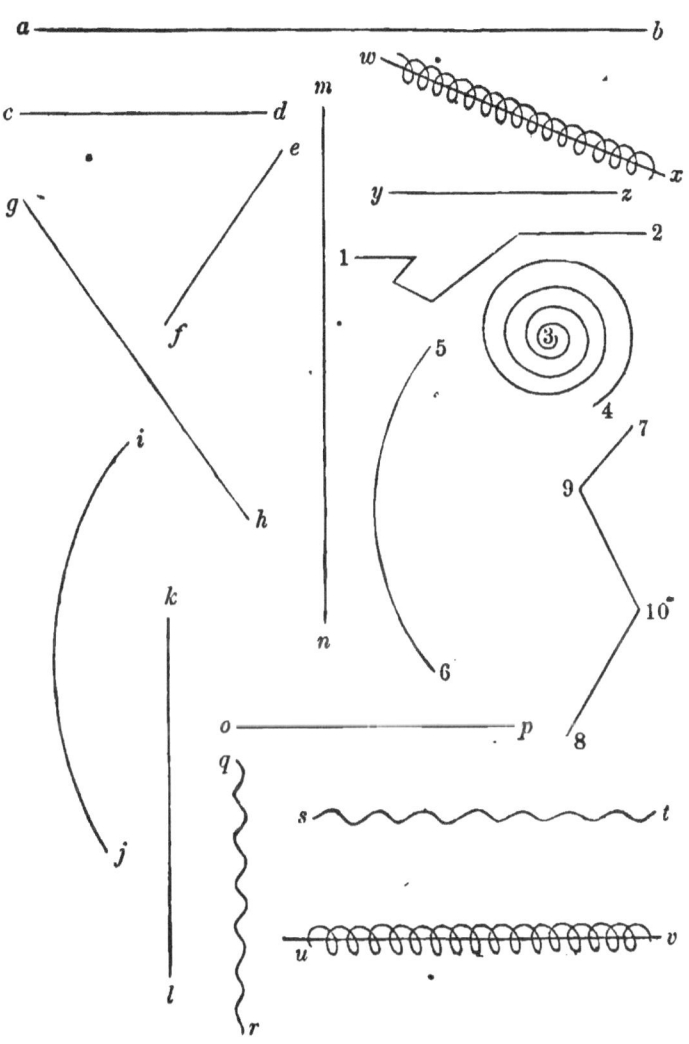

Diagram 2.

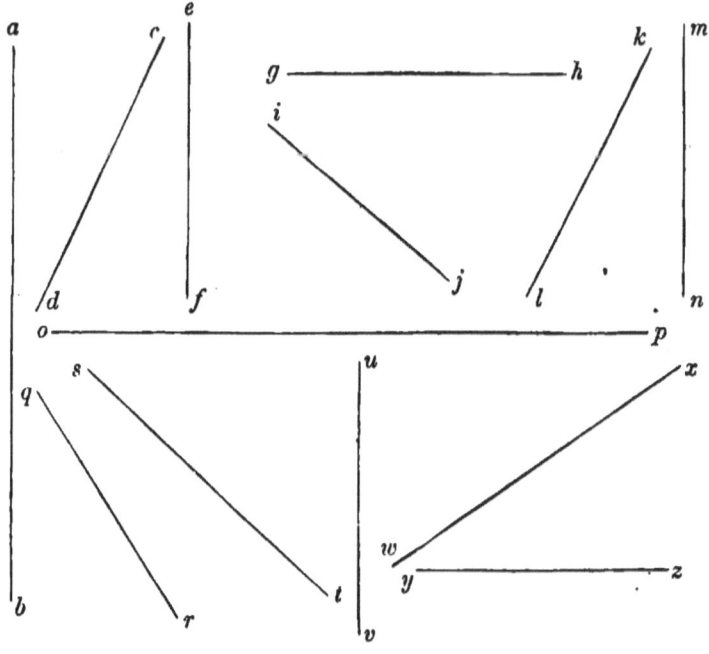

Diagram 3.

LESSON THIRD.

POSITIONS OF LINES.

Let the pupils hold their books so that they will be straight up and down like the wall.

VERTICAL LINES.

The straight line *a b* points to the center of the earth. (DIAGRAM 3.)

It is called a vertical line.

Name all the vertical lines.

A vertical line is a straight line that points to the center of the earth.

HORIZONTAL LINES.

The straight line *o p* points to the horizon.

It is called a horizontal line.
Read all the horizontal lines.

A horizontal line is a straight line that points to the horizon.

OBLIQUE LINES.

The line *s t* points neither to the center of the earth nor to the horizon.
It is called an oblique line.
Read all the oblique lines.

An oblique line is a straight line that points neither to the horizon nor to the center of the earth.

NOTE.— After going through with the lessons on angles, the pupils may be told that oblique lines are so called because they form oblique angles with the horizon.

LESSON FOURTH.

REVIEW.

Read all the vertical lines. (DIAGRAM 4.)
Why is *q r* a vertical line ?
What is a vertical line ?
Read all the horizontal lines.
Why is 5 6 a horizontal line ?
Define a horizontal line.
Read all the oblique lines.
Why is *s t* an oblique line.
What is an oblique line ?

NOTE. — Lines that point in the same direction do not approach the same point.

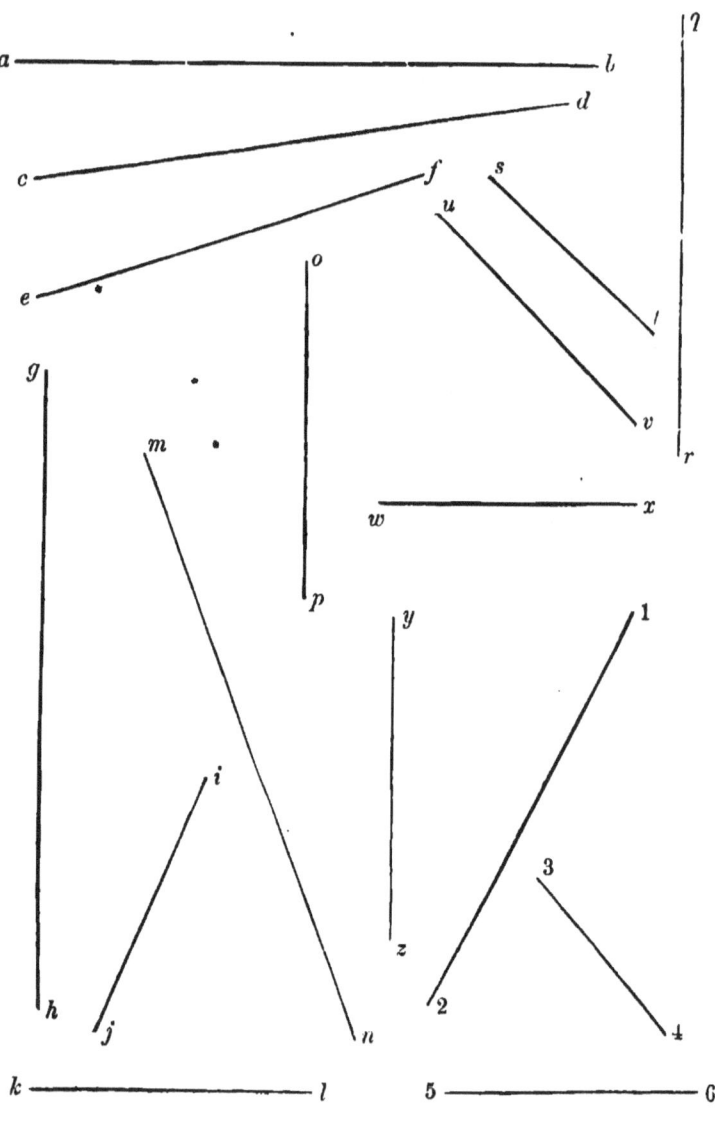

Diagram 4.

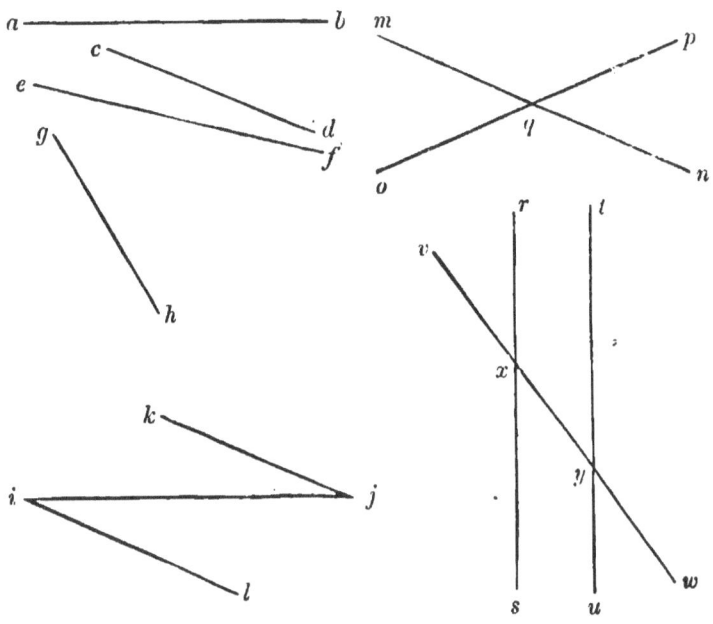

Diagram 5.

LESSON FIFTH.

ANGLES.

Do the lines *a b* and *c d* (DIAGRAM 5) point in the
same direction?   (See note, page 15.)

Then they form an *angle* with each other.

What other line forms an angle with *a b ?*

Which of the two lines *c d, e f,* has the greater difference of direction from the line *a b ?*

Then which one forms the greater angle with *a b ?*

What line forms a still greater angle with the line
a b ?

An angle is the difference of direction of two straight lines.

If the lines *a b*, *e f*, were made longer, would their direction be changed?

Then would there be any greater or less difference of direction?

Then would the angles formed by them be any greater or less?

Then does the *size* of an angle depend upon the length of the lines that form it?

If the lines *a b*, *e f*, were shortened, would the angle formed by them be any smaller?

If two lines form an angle with each other, and meet, the point of meeting is called the vertex.

What is the vertex of the angle formed by the lines *k j*, *i j* ? — *i j*, *i l* ?

An angle is named by three letters, that which denotes the vertex being in the middle. Thus, the angle formed by *k j*, *i j*, is read *k j i*, or *i j k*.

Read the four angles formed by the lines *m n* and *o p*.

The eight formed by *r s*, *t u*, and *v w*.

LESSON SIXTH.

REVIEW.

Read all the lines that form angles with the line *a b*. (DIAGRAM 6.)

Which of them forms the greatest angle with it?

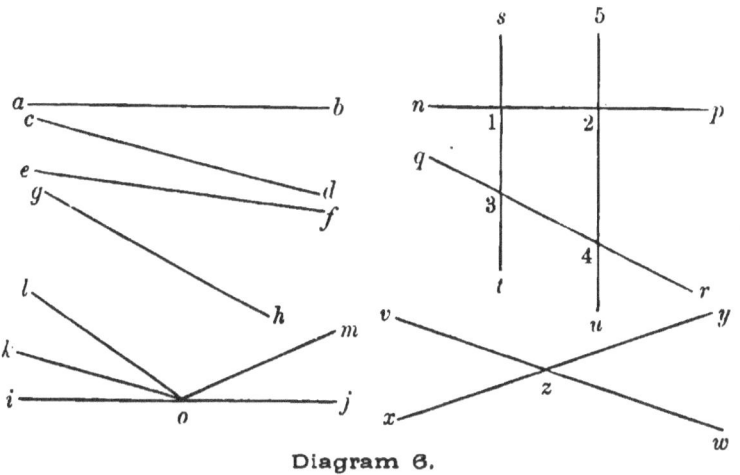

Diagram 6.

Which the least?

Of the two lines *c d, g h,* which forms the greater angle with *e f?*

Read all the angles whose vertices are at *o* on *i j.*

Which angle is the greater, *l o m,* or *m o j?* — *i o k,* or *i o l?* — *l o j,* or *m o j?*

Read all the angles formed by the lines *v w* and *x y.*

Read all the angles above the line *n p.*

Below the line *n p.*   Above the line *q r.*

At the right of the line 5 *u.*

At the left. *At the right of the line *s t.*

At the left of the line *s t.*

Which angle is the greater, *n* 1 3, or *n* 2 4?

If the lines *x y* and *v w* were lengthened or produced, would the angles *v z x, y z w* be any greater?

If they were shortened, would the angles be any less?

What is an angle?
Does the size of an angle depend upon the length
of the lines which form it?

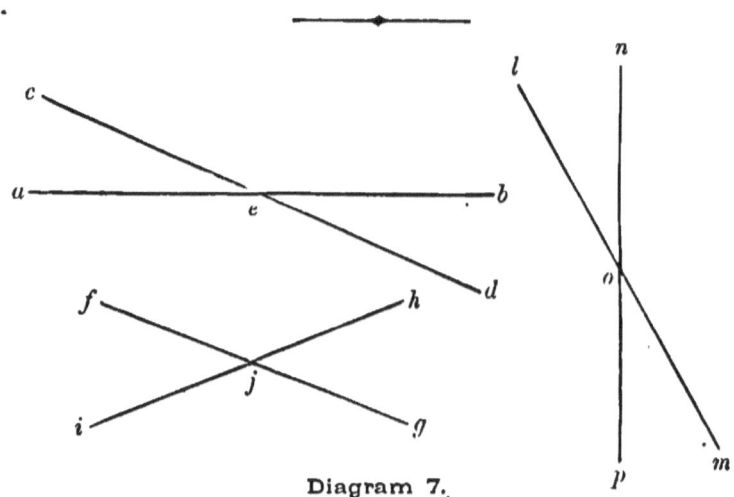

Diagram 7.

LESSON SEVENTH.

RELATIONS OF ANGLES.

ADJACENT ANGLES.

Are the angles $a\,e\,c,\,c\,e\,b$ (DIAGRAM 7), on the same
side of any line? What line?
By what other straight line are they both formed?
Then, because they are both on the same side of
the same straight line $a\,b$, and are both formed
by the second straight line $c\,d$, they are called
"*adjacent angles*."
The angles $c\,e\,b,\,b\,e\,d$ are both on the same side
of what straight line?

They are both formed by what second straight line?

Then what kind of angles are they?

Why are they called adjacent angles?

Read the adjacent angles below the line *a b*. Below the line *c d*.

How many pairs of adjacent angles can be formed by two straight lines?

Read all the adjacent angles formed by the lines *l m* and *n p*.

VERTICAL ANGLES,

Are the angles *a e c*, *b e d* formed by the same straight lines?

Are they adjacent angles?

They are called "vertical angles."

Vertical angles are angles formed by the same straight lines, but not adjacent to each other.

Read the other pair of vertical angles formed by the lines *a b*, *c d*.

Read all the vertical angles formed by the lines *f g*, *i h*. By *l m*, *n p*.

Why are the angles *l o n*, *n o m* adjacent angles?

Why are the angles *l o n*, *p o m* vertical angles?

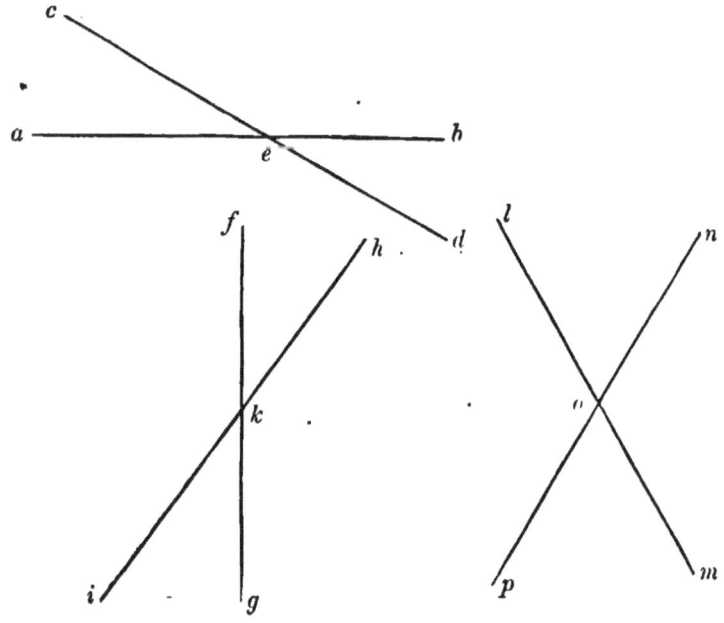

Diagram 8.

LESSON EIGHTH.

REVIEW.

Read the pairs of adjacent angles above the line
a b. (Diagram 8.)

Why are they adjacent?

What are adjacent angles?

Read the adjacent angles below the line *a b*.

On the right of the line *c d*. On the left.

How many pairs of adjacent angles are formed by
the intersection of two lines.

Read the pairs of adjacent angles formed by the
lines *f g* and *i h*.

Read all the adjacent angles formed by the lines
l m, n p.

Read all the pairs of vertical angles formed by the
lines *a b, c d.*

Why are *c e b* and *a e d* called vertical angles?

What are vertical angles?

Read all the pairs of vertical angles formed by the
lines *h i, f g.*

How many pairs of vertical angles are formed by
the intersection of two lines?

Read all the pairs of vertical angles formed by the
lines *l m, n p.*

LESSON NINTH.

KINDS OF ANGLES.

RIGHT ANGLES.

What do we call the angles *a o c, c o b?* (DIAGRAM 9.)
Are they equal to each other?

Then they are called *right angles.*

*A right angle is one of two adjacent angles that are
equal to each other.*

Are the adjacent angles. *c o b, b o d* equal to each
other?

Then what are they called?

Read the right angles below the line *a b.* On the
left of *c d.*

Read three right angles whose vertices are at *p.*

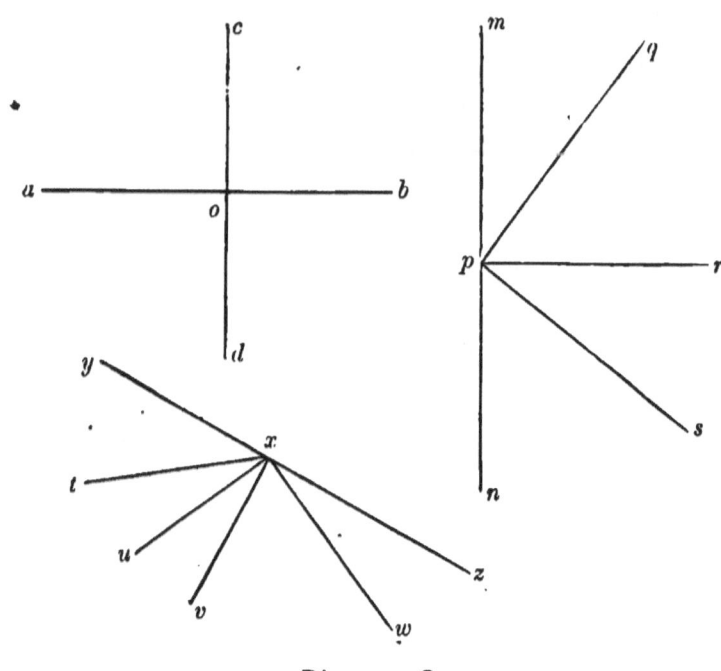

Diagram 9.

ACUTE ANGLES.

Is the angle $m\,p\,q$ greater or less than the right
angle $m\,p\,r$?

Then it is called an *acute angle*.

An acute angle is one which is less than a right angle.

Read four acute angles whose vertices are at p.
Acute means sharp.

Why is $r\,p\,s$ an acute angle?

What is an acute angle?

OBTUSE ANGLES.

Is the angle $m\,p\,s$ greater or less than the right
angle $m\,p\,r$?

Then it is called an *obtuse angle*.

An obtuse angle is one which is greater than a right angle.

What other obtuse angle has its vertex at *p ?*

Obtuse means blunt.

Read three obtuse angles whose vertices are at *x*.

Acute and obtuse angles are also called oblique angles.

LESSON TENTH.

REVIEW.

Read all the right angles formed by the lines *a b* and *c d*. (DIAGRAM 10.)

Why are the adjacent angles *c e b*, *b e d*, right angles?

What is a right angle?

Read four right angles whose vertices are at *n*.

Which is the greater, the right angle *p q r*, or the right angle *t s u ?*

Can one right angle be greater than another?

Read six acute angles whose vertices are at *n*.

Why is *m n g* an acute angle?

What is an acute angle?

Which is greater, the acute angle *m n g*, or the acute angle *l n m ?*

May one acute angle be greater than another?

What three acute angles are equal to one right angle?

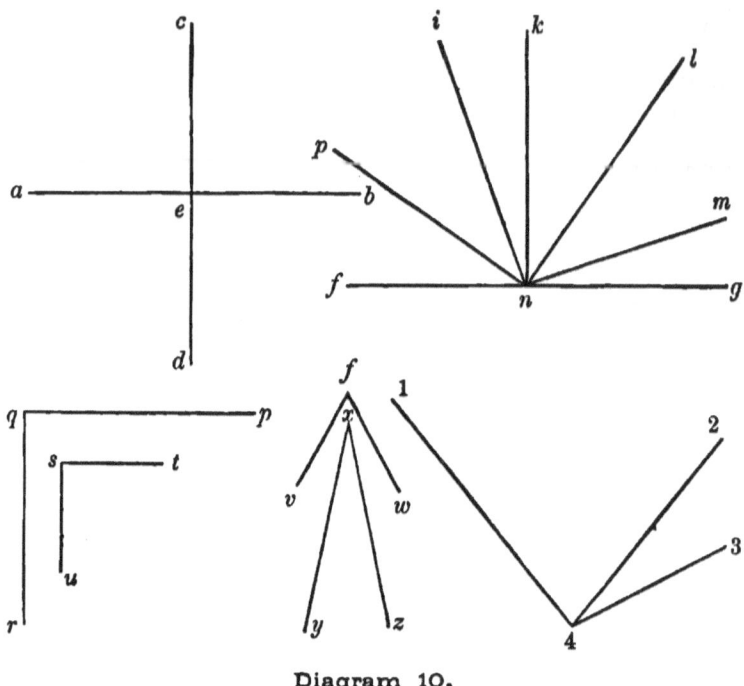

Diagram 10.

Which of the two acute angles $v f w$, $y x z$ is the greater?

Read four obtuse angles whose vertices are at n.

Why is $f n m$ an obtuse angle?

What is an obtuse angle?

What does obtuse mean?   Acute?

By what other name are both called?

Which is greater, the large acute angle 1 4 2, or the small obtuse angle 1 4 3?

How much greater than the right angle is the obtuse angle $f n l$?

How much less than a right angle is $f n i$?

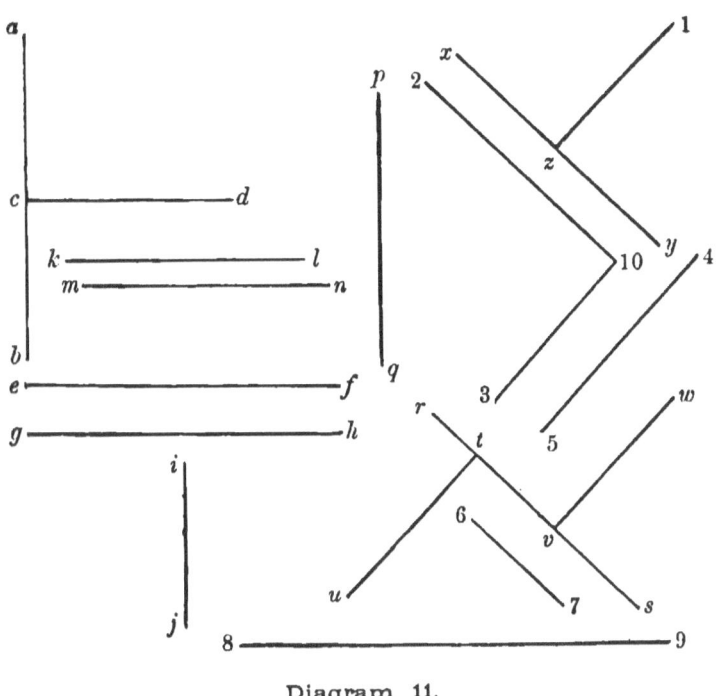

Diagram 11.

LESSON ELEVENTH.

RELATIONS OF LINES.

PERPENDICULAR LINES.

What kind of angles do the lines $a\,b$ and $c\,d$ make
with each other? (DIAGRAM 11.)

Then they are perpendicular to each other.

What line is perpendicular to $x\,y$?

Why is it perpendicular to it?

What line is perpendicular to $z\,1$?

When is a line said to be perpendicular to another?

Can a line standing alone be properly called a perpendicular line?

What two lines are perpendicular to the lines *r s?*

Is the line *g h* perpendicular to the line *i j?* Why?

What other line is perpendicular to the line *i j?*

Read three lines that are perpendicular to the line *a b.*

PARALLEL LINES.

Do the lines *k l, m n,* differ in direction? Then do they form any angle with each other?

They are said to be *parallel* to each other.

Read four other lines that are parallel with *k l.*

What line is parallel with 2 10?

Why?

Lines are parallel with each other when they do not differ in direction.

OBLIQUE LINES.

What kind of angles do the lines *u t* and 8 9 form with each other?

Then they are said to be oblique to each other.

Lines are oblique to each other when they form oblique angles.

See Note C, Appendix.

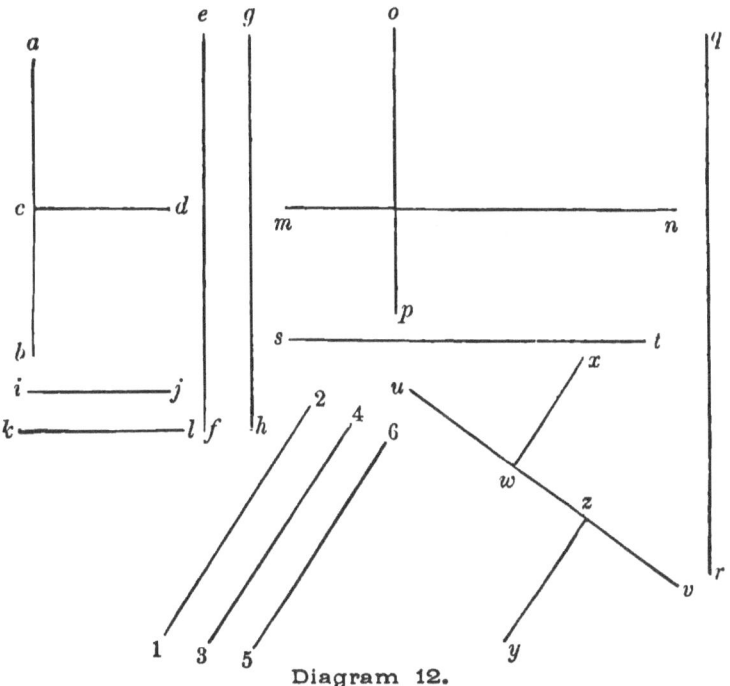

Diagram 12.

LESSON TWELFTH.

REVIEW.

Read five lines that are perpendicular to the line
a b. (DIAGRAM 12.)

Five that are perpendicular to *c d*.

Two that are perpendicular to *u v*, and meet it.
Three that do not meet it.

Why are *o p* and *m n* perpendicular to each other?

When are lines said to be perpendicular to each
other?

Read four lines that are parallel with *e f*.

Why are the lines *e f* and *g h* said to be parallel to
each other?

When are lines said to be parallel to each other?

Read four lines that are parallel to 5 6.

Four that are parallel to *o p*.

Is any line parallel to *u v?*

Can a single line be properly called perpendicular?
Parallel?

If two lines are perpendicular to each other, what
angle do they form?

If parallel, what angle?   If oblique?

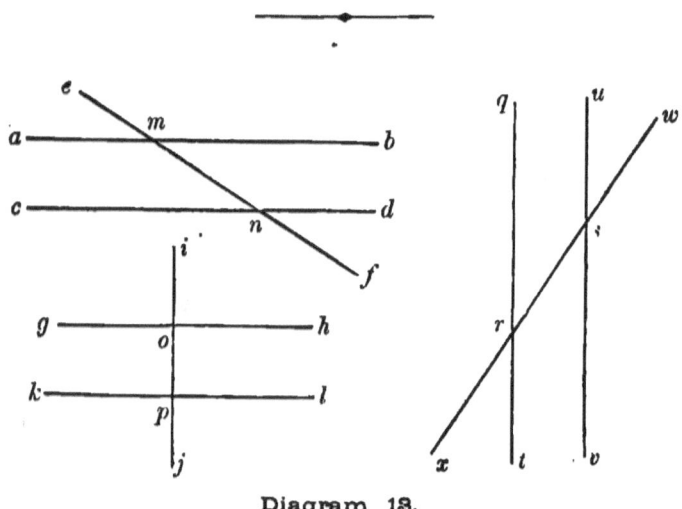

Diagram 13.

LESSON THIRTEENTH.

RELATIONS OF ANGLES.

INTERIOR ANGLES.

Is the angle *a m n* between the parallels, or outside
of them?   (DIAGRAM 13.)

It is called an *interior angle.*

Read three other interior angles between the same parallels.

Why is *b m n* an interior angle?

An interior angle is one that lies between parallel lines.

Read the interior angles between the parallel lines *g h* and *k l.*

Why is *o p l* an interior angle?

What is an interior angle?

EXTERIOR ANGLES.

Is the angle *a m e* between the parallels, or outside of them?

It is called an *exterior angle.*

Read three other exterior angles formed by the lines *a b, c d,* and *e f.*

Why is the angle *c n f* an exterior angle?

An exterior angle is one that lies outside of the parallels.

LESSON FOURTEENTH.

REVIEW.

Read all the interior angles formed by the lines *a b, c d,* and *e f.*

Why is *m n d* an interior angle?

What is an interior angle?

Read all the exterior angles formed by the same lines.

Why is *d n f* an exterior angle?

What is an exterior angle?

Read all interior angles formed by the lines *g h, k l,*
   and *i j.*

All the remaining interior angles in the diagram.
   All the exterior angles.

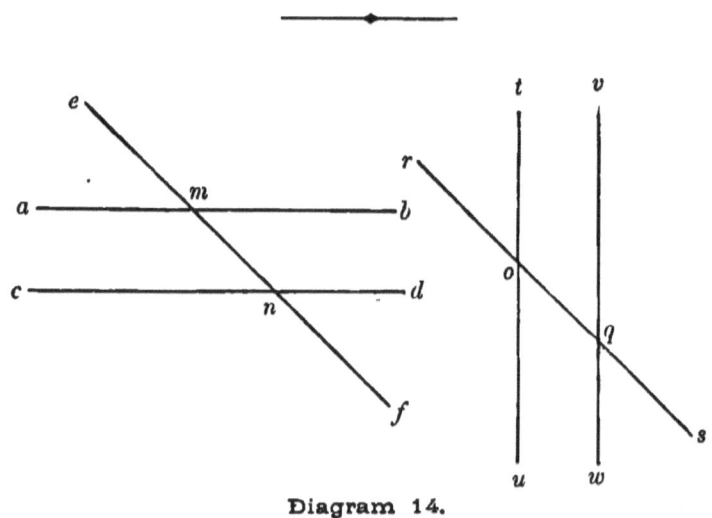

Diagram 14.

LESSON FIFTEENTH.

RELATIONS OF ANGLES.

OPPOSITE ANGLES.

Are the angles *e m b, b m n,* on the same side of the
   intersecting line *e f ?*

Are they adjacent?

Are *e m b, m n d,* on the same side of the intersecting line *e f ?*

Are they adjacent?

Then they are called opposite angles.

Opposite angles lie on the same side of the intersecting line, but are not adjacent.

Are the angles *e m b, f n d,* on the same side of the intersecting line?
Are they adjacent?
Then are they opposite?
Are they interior or exterior angles?
Then they are "*opposite exterior angles.*"
Why are they exterior?
Why are they opposite?
Are the angles *b m n, m n d,* opposite angles?
Are they interior or exterior angles?
Then they are "*opposite interior angles.*"
Why are they opposite? Why interior?
Read the opposite exterior angles on the left of the line *e f.*
Read the opposite interior angles on the same side.
Are the opposite angles *e m a, m n c,* both exterior or interior?
Then they are *opposite exterior and interior angles.*
Read two pairs of opposite exterior and interior angles on the right of *e f.* On the left.

ALTERNATE ANGLES.

Do the angles *b m n, m n c,* lie on the same side of the intersecting line *e f?*
Are they adjacent to each other?
Are they vertical angles?
Then they are alternate angles.

Alternate angles lie on different sides of the intersect-
ing line, and are neither adjacent nor vertical.

Are the alternate angles *b m n, m n c,* exterior or
   interior?

Then they are called "*interior alternate angles.*"

Read another pair of interior alternate angles
   between *a b* and *c d.*

Are the angles *e m b, c n f,* alternate angles? Why?

Are they exterior or interior?

Then what may they be called?

Read another pair of exterior alternate angles.

Why are *e m a, d n f,* alternate angles? Why exterior alternate?

LESSON SIXTEENTH.

REVIEW.

Read the exterior opposite angles on the right of
the line *e f*. (DIAGRAM 14.)

On the left. On the right of *r s*. On the left.

Why are *e m a, c n f*, exterior angles?

Why are they opposite angles?

What are opposite angles?

Read the interior opposite angles on the right of
the intersecting line *c f*.

On the left of it. On the right of *r s*. On the left.

Read the interior alternate angles formed by the
lines *a b, c d*, and *e f*.

Which pair are acute angles?

Which pair are obtuse angles?

Why are *b m n, m n c*, interior angles? Why alternate? What are alternate angles?

Read the exterior alternate angles of the same
lines.

Read the acute interior alternate angles of the
parallels *t u, v w*. The obtuse.

The acute exterior alternate angles. Obtuse.

Read the pair of opposite exterior angles on the
right of the line *e f*. On the left.

On the right of *r s*. On the left.

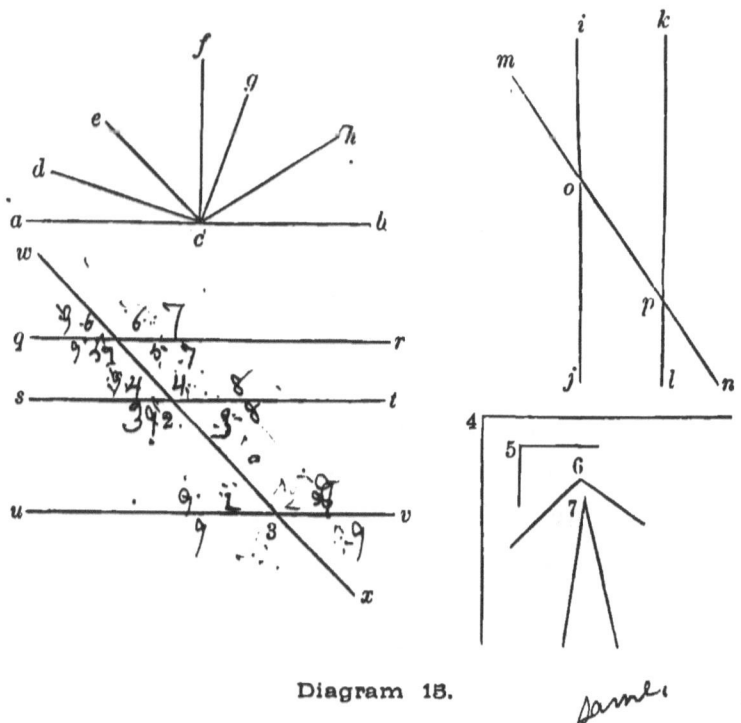

Diagram 15.

LESSON SEVENTEENTH.

REVIEW.

Read thirteen or more angles whose vertices are at *c*. (Diagram 15.)

Read four obtuse angles.

Read two right angles.

What three acute angles equal one right angle?

Which is greater, the right angle 4, or the right angle 5?

The obtuse angle 6, or the acute angle 7?.

Read twelve pairs of adjacent angles formed by the lines *w x*, &c.

Read six pairs of vertical angles formed by the same lines.

Read all the interior angles formed by the lines *i j*, *k l*, and *m n*.

Read all the exterior angles formed by the same lines.

Two pairs of opposite exterior angles.

Two pairs of opposite interior angles.

Four pairs of opposite exterior and interior angles.

Two pairs of alternate interior angles.

Two pairs of alternate exterior angles.

Why are the angles *i o m*, *m o j*, called adjacent?

What are adjacent angles?

What kind of an angle is *i o m?*   Why?

What is an acute angle?

What kind of an angle is *m o j?*   Why?

What is an obtuse angle?

Why are *a c f*, *f c b*, right angles?

What is a right angle?

Why are *m o i*, *j o p*, vertical angles?

What are vertical angles?

Why is *m o i* an exterior angle?

What is an exterior angle?

Why is *j o p* an interior angle?

What is an interior angle?

Why are *m o i*, *o p k*, opposite angles?

What are opposite angles?

Why are *j o p*, *o p k*, alternate angles?

What are alternate angles?

LESSON EIGHTEENTH.

PROBLEMS.

Draw an obtuse angle which shall be only a little larger than a right angle.

Draw one which shall be much greater than a right angle.

Draw an acute angle which shall be only a little less than a right angle.

Draw one which shall be much less than a right angle.

Draw an obtuse angle with lines about one inch long.

Draw an acute angle with sides three inches long.

Which is greater, the obtuse angle, or the acute angle?

Draw a right angle with lines an inch long.

Draw one with lines five inches long.

Which is the greater, first or the second?

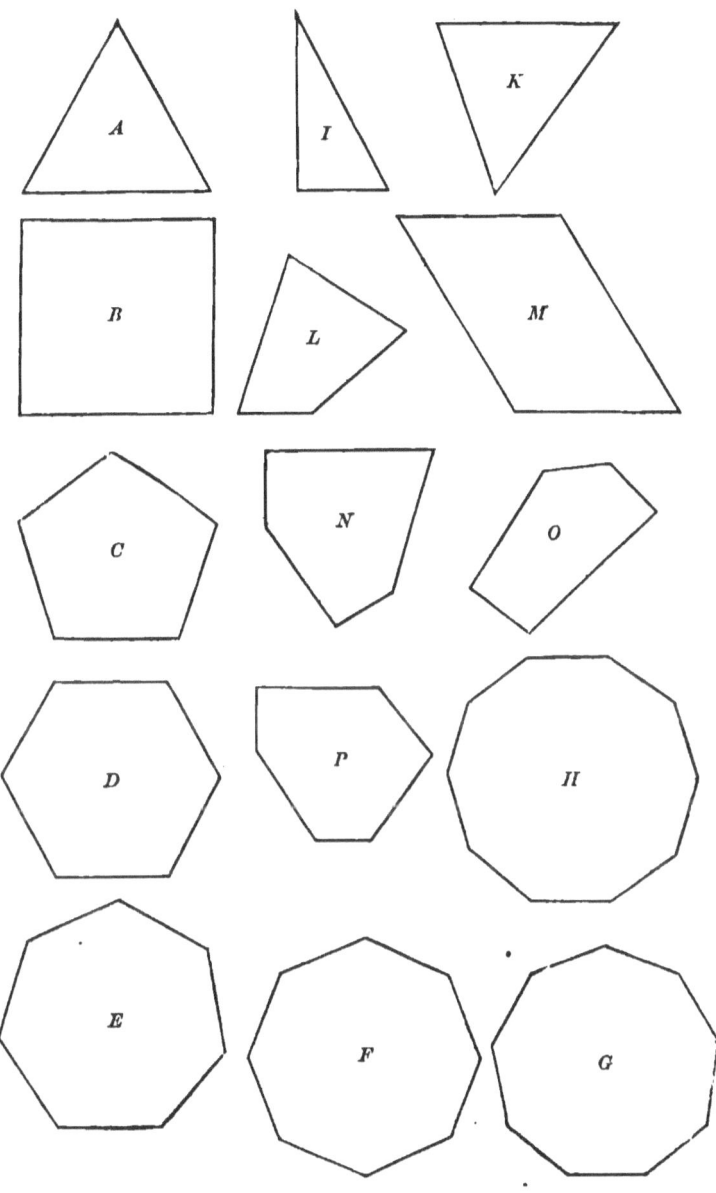

Diagram 16.

LESSON NINETEENTH.

POLYGONS.

Name any thing besides your desk that has a flat
surface.

A flat surface is called a plane.

How many sides has the plane Fig. A ? (DIAGRAM
16.)

It is called a triangle. " Tri " means " three."

What other triangles do you see.

Triangles are sometimes called trigons.

A triangle is a plane figure having three sides.

How many sides has the plane figure marked B ?
How many angles ?

It is called a quadrangle, or quadrilateral. " Quad "
denotes " four."

What other quadrangles do you see ?

Why is Fig. B a quadrangle ?

A quadrangle is a plane figure having four sides.

How many sides has the Fig. C ?

It is called a *pentagon.*

What other pentagon do you see ?

Why is Fig. C a pentagon ?

A pentagon is a plane figure having five sides.

In like manner, —

A hexagon is a plane figure having six sides.

A heptagon is a plane figure having seven sides.

An octagon has eight sides.

A nonagon has nine sides.

A decagon has ten sides.

All these figures are called *polygons*.

" Poly " means " many."

What do you call a polygon of three sides? Of four sides? Of six sides? &c.

If the length of each side of triangle A is one inch, how long are the three sides together?

The sum of the sides of a polygon is its perimeter.

Which of the triangles has unequal sides? Which has equal sides?

The latter is called a *regular polygon*.

Which pentagon has one side longer than any one of its other sides?

Which has its sides all equal to each other? Are its angles also equal?

It is therefore a *regular polygon*, or *regular pentagon*.

Name a hexagon that is not regular.

Name a regular hexagon.

A regular octagon. A regular heptagon.

A polygon is a plane figure bounded by straight lines.

LESSON TWENTIETH.

REVIEW.

Name all the triangles.   (DIAGRAM 16.)

Why is Fig. A a triangle?

What is a triangle?

What other name is sometimes given to triangles?

Name all the quadrilaterals.

Why is Fig. B a quadrilateral?

What is a quadrilateral, or quadrangle?

Name all the pentagons, hexagons, heptagons, octagons, and nonagons.

Why is C a pentagon?   What is a pentagon?   A hexagon?   A heptagon? &c.

How many polygons in the diagram?

What is a polygon?   .

If each side of Fig. B is one inch, how many inches are there in its perimeter?

When is a polygon regular?

Name all the regular polygons in diagram 16.

Name all the irregular polygons.

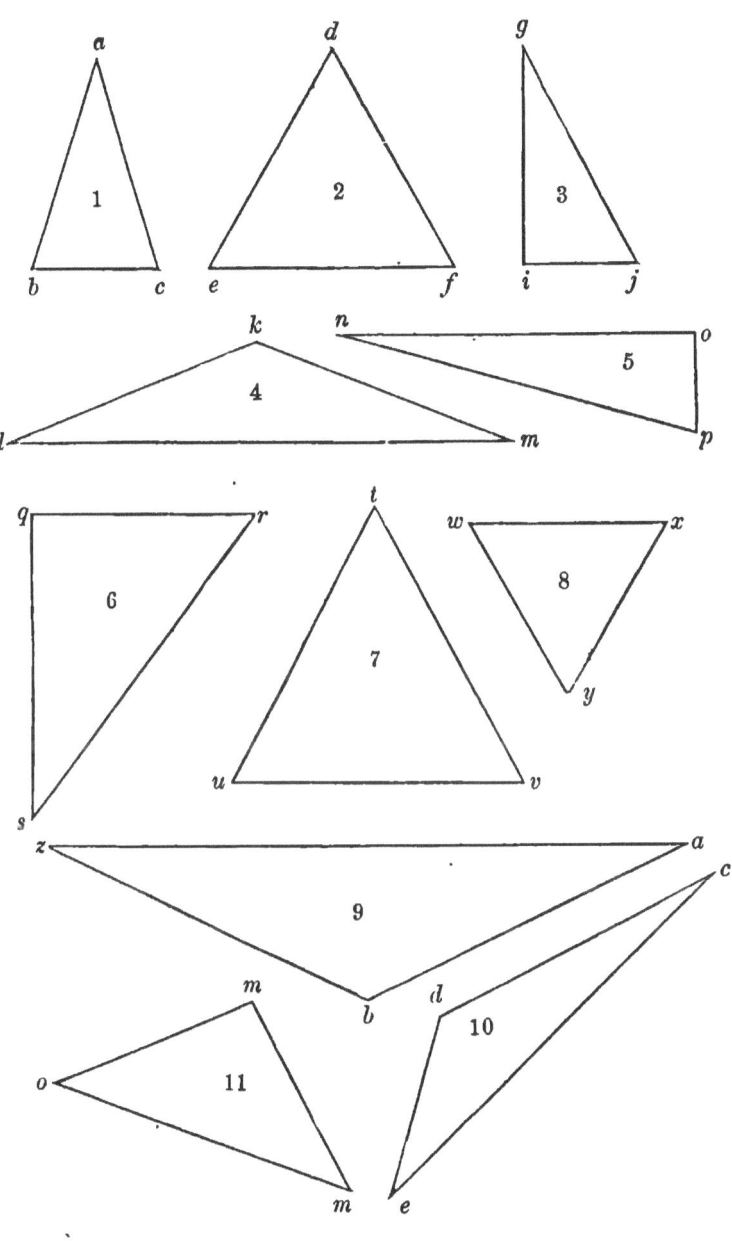

Diagram 17.

LESSON TWENTY-FIRST.

TRIANGLES.

ACUTE-ANGLED TRIANGLES.

In the triangle 1, what kind of an angle is $b\ a\ c$? $a\ c\ b$? $c\ b\ a$? (DIAGRAM 17.)

Then it is called an *acute-angled triangle*.

An acute-angled triangle is one whose angles are all acute.

Read three other acute-angled triangles,

OBTUSE-ANGLED TRIANGLES.

In the triangle 4, what kind of an angle is $l\ k\ m$?

Then it is called an *obtuse-angled triangle*.

An obtuse-angled triangle is one that has one obtuse angle.

Name two others.

RIGHT-ANGLED TRIANGLES.

In the triangle 3, what kind of an angle is $g\ i\ j$?

Then it is called a *right-angled triangle*.

A right-angled triangle is one that has one right angle.

Name three other right-angled triangles.

Upon which side does the triangle 3 seem to stand?

Then $i\ j$ is called the *base* of the triangle.

What letter marks the vertex of the angle opposite the base?

Then the point g is said to be the vertex of the triangle.

If, in the triangle 7, we consider $t\,v$ the base, what point is the vertex?

If v be considered the vertex, which side will be the base?

In the triangle 3, what side is opposite the right angle?

Then $g\,j$ is called the *hypothenuse* of the triangle.

The hypothenuse of a triangle is the side opposite the right angle.

Read the hypothenuse of each of the triangles 5, 6, and 11.

Either side about the right angle may be considered the base.

Then the other side will be the perpendicular.

In the triangle 3, if $i\,j$ is the base, which side is the perpendicular?

If $g\,i$ be considered the base, which side is the perpendicular?

In triangle 5, if $n\,o$ is the base, which side is the perpendicular?

LESSON TWENTY-SECOND.

REVIEW.

Name four acute-angled triangles.   (DIAGRAM 17.)

Why is the triangle 8 acute-angled?

What is an acute-angled triangle?

Name three obtuse-angled triangles.

Why is the triangle 9 an obtuse-angled triangle?

What is an obtuse-angled triangle?

Name four right-angled triangles.

Why is the triangle 6 a right-angled triangle?

What is a right-angled triangle?

In the triangle 6, which side is the hypothenuse?
Why?

What is the hypothenuse?

What two sides of the triangle 6 may be regarded
as the base?

If $q\,r$ be considered the base, what do you call the
side $q\,s$?

Read the hypothenuse of each of the triangles 3
5, 6, and 11.

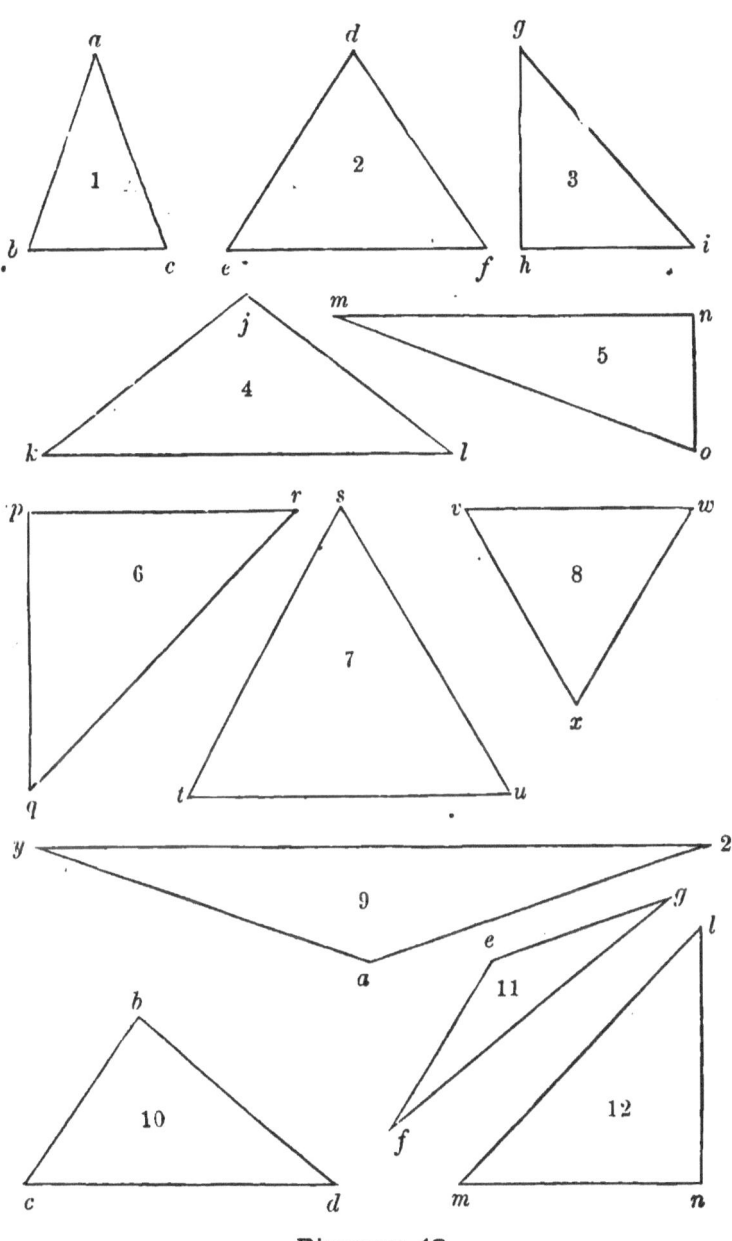

Diagram 18.

LESSON TWENTY-THIRD.

TRIANGLES. (*Continued.*)

ISOSCELES TRIANGLES.

Of the triangle 1, which two sides are equal to
each other?

Then it is called an *isosceles triangle*.

An isosceles triangle is one that has two equal sides.

Name eight isosceles triangles.

Why is the triangle 2 an isosceles triangle?

What kind of a triangle is it on account of its
angles?

Then it is an *acute-angled isosceles triangle*.

Name four acute-angled isosceles triangles.

What kind of a triangle is Fig. 4 on account of the
angle *k j l?*

What kind on account of its equal sides?

Then it is called an *obtuse-angled isosceles triangle*.

Name one other obtuse-angled isosceles triangle.

What kind of a triangle is Fig. 6 on account of the
angle *q p r?*

What kind on account of its equal sides?

Then it is called a *right-angled isosceles triangle*.

Name one other right-angled isosceles triangle.

Why is Fig. 12 a right-angled triangle? Why
isosceles?

EQUILATERAL TRIANGLES.

Which of the isosceles triangles has all its three
sides equal to each other?

It is called an *equilateral triangle.*

" Equi " means " equal."   " Latus " means a " side."

An equilateral triangle is one that has its three sides
equal to each other.

What kind of a triangle is Fig. 7 on account of its
three equal sides?

What kind on account of its two equal sides *s t, s u,*
or *t s, t u,* or *u s, u t?*

Then must not every equilateral triangle be also
isosceles?

What kind of a triangle is Fig. 2 on account of its
equal sides *d e, d f?*

If the side *e f* is longer than either of the other
two sides, is it an equilateral triangle?

Then is every isosceles triangle also equilateral?

Name another isosceles triangle that is *not* equilateral.

Name one that *is* equilateral.

In any equilateral triangle the three angles are
equal to each other.

On account of its equal angles, it is also called
an *equiangular triangle.*

What is Fig. 8 called on account of its three equal
sides?   On account of its three equal angles?

Every equilateral triangle is also equiangular.

Every equiangular triangle is also equilateral.

Name a triangle that has no two sides equal to
each other.

It is called a *scalene triangle*.

What kind of a triangle is Fig. 5 on account of its
right angle?

What kind on account of its three unequal sides?

Then it is a *right-angled scalene triangle*.

What name can you give Fig. 11 on account of the
angle *g e f?*

On account of its three unequal sides?

Then what may it be called?

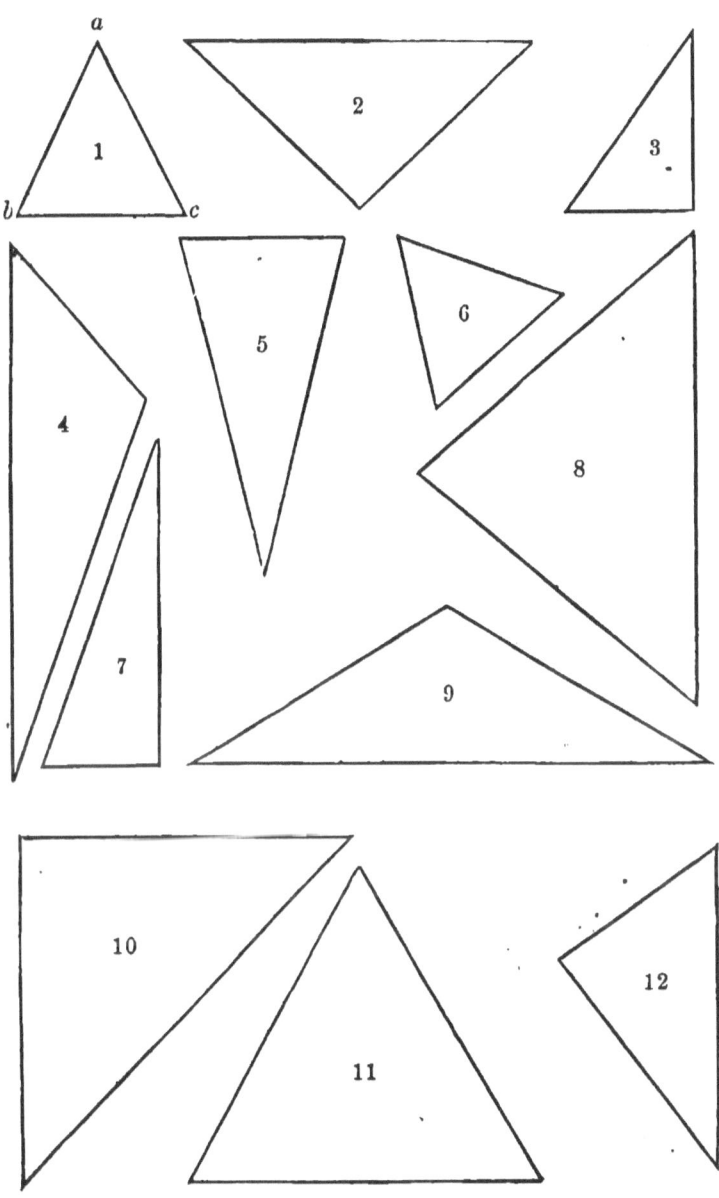

Diagram 19.

LESSON TWENTY-FOURTH.

REVIEW.

Name eight isosceles triangles. (DIAGRAM 19.)

Why is Fig. 2 an isosceles triangle?

What is an isosceles triangle?

Name two right-angled isosceles triangles.

Name five acute-angled isosceles triangles.

Name one obtuse-angled isosceles triangle.

Name two isosceles triangles that are also equilateral.

Are all isosceles triangles equilateral?

Name six isosceles triangles that are *not* equilateral.

What does " equi" mean?   " Latus"?

What are equilateral triangles called on account of their equal angles?

Are all equilateral triangles equiangular?

Are all equiangular triangles equilateral?

What are equilateral triangles?

Name four scalene triangles.

Name two right-angled scalene triangles.

Why is Fig. 3 a right-angled triangle?   Why scalene?

What is a scalene triangle?

Name one obtuse-angled scalene triangle.

Name one acute-angled scalene triangle.

PROBLEMS.

From the same point draw two straight lines of any length, making an acute angle with each other.

Make them equal to each other by measuring.

Join their ends.

What kind of a triangle is it on account of its angles?

On account of its two equal sides?

Write its two names inside of it.

Draw an isosceles triangle whose equal sides shall each be less than the third side.

Write its two names within it.

Draw an oblique straight line twice as long as any short measure or unit.

At one end draw a straight line perpendicular to it, and three times as long as the same measure.

Connect the ends of the two lines by a straight line.

What kind of an angle is that opposite the last line drawn?

Are any two of its sides equal?

Write its two names under it.

Draw a horizontal straight line of any length.

At one end draw a vertical line of equal length.

Complete the triangle, and write two names inside.

Draw a right-angled triangle whose base is of any length, and its perpendicular twice as long.

Draw a right-angled triangle whose base is three times as long as any short measure, and its perpendicular five times as long as the same measure or unit.

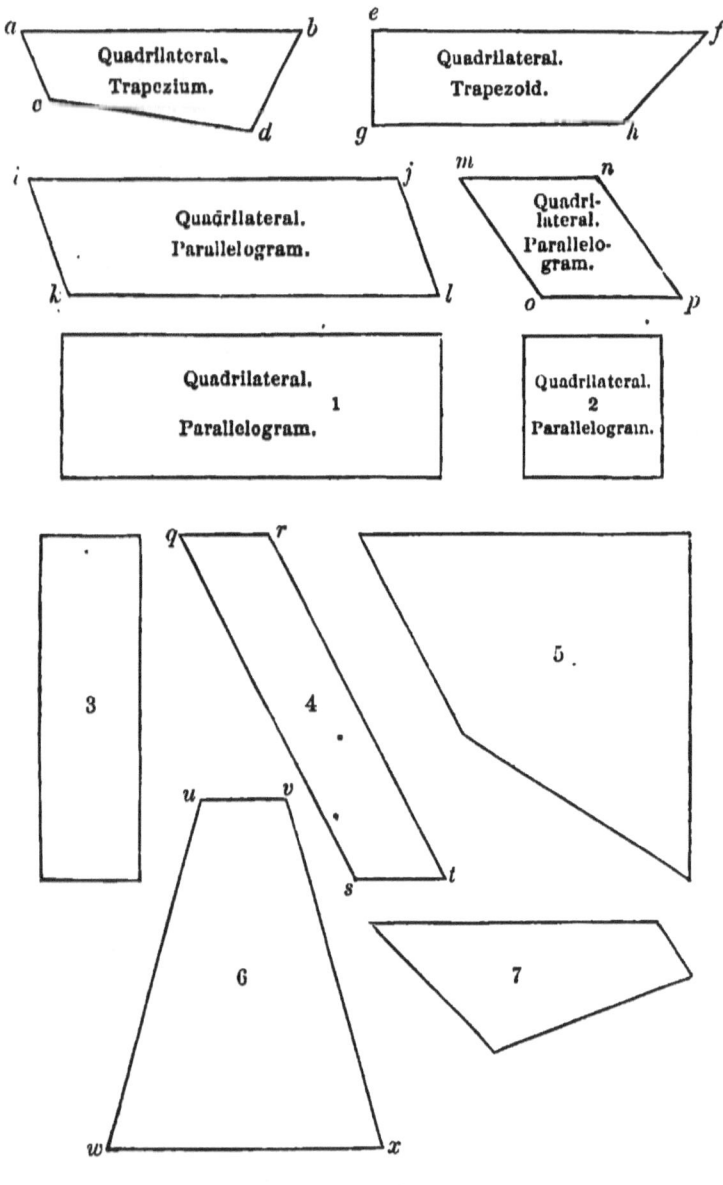

a   b

Quadrilateral.
Trapezium.

c

d

e   f

Quadrilateral.
Trapezoid.

g   h

i   j

Quadrilateral.
Parallelogram.

k   l

m   n

Quadri-
lateral.
Parallelo-
gram.

o   p

Quadrilateral.
1
Parallelogram.

Quadrilateral.
2
Parallelogram.

q   r

3

4

5

u   v

s   t

6

7

w   x

Diagram 20.

3

QUADRILATERALS.

How many sides has the figure $a\ b\ d\ c$?

What is it called on account of the number of its sides?

Name three other quadrilaterals whose vertices are marked.

Name seven by numbers.

Quadrilaterals are sometimes named by means of two opposite vertices.

The quadrilateral $a\ b\ d\ c$, or $c\ d\ b\ a$, may be read $a\ d$, or $b\ c$, or $c\ b$, or $d\ a$.

Name the quadrilateral, $g\ h\ f\ e$, four ways.

How many angles has each figure?

On account of the number of their angles they are called *quadrangles*.

Has the quadrilateral $a\ d$ any two sides parallel to each other?

Then it is called a *trapezium*.

A trapezium is a quadrilateral that has no two sides parallel.

Name two other trapeziums.

Why is Fig. 7 a trapezium?

Has the quadrilateral $e\ h$ any two sides parallel? Which two? Are the other two sides parallel?

It is called a " *trapezoid*."

" Oid " means like. What does " trapezoid " mean ?

A trapezoid is a quadrilateral that has only one pair of sides parallel.

Name another trapezoid.

Why is Fig. 6 a trapezoid?

How many pairs of parallel sides has the quadrilateral *i l?*

Name the horizontal parallels.

Name the oblique parallels.

It is called a "*parallelogram.*"

*A parallelogram is a quadrilateral whose opposite
sides are parallel.*

Name five other parallelograms.

Why is Fig. 4 a parallelogram?

Why is not Fig. 6 a parallelogram?

Why is not *e h* a parallelogram?

What two names may you give to Fig. 5?

Why is it a quadrilateral?   Why a trapezium?

What two names may we give to Fig. 6?

Why is it a quadrilateral?   Why a trapezoid?

What two names may we give to Fig. 3?

Why is it a parallelogram?   Why a quadrilateral?

LESSON TWENTY-FIFTH.

REVIEW.

How many quadrilaterals in the diagram. (DIA-
GRAM 20.)

Why is Fig. *a d* a quadrilateral?

What is a quadrilateral?

On account of the number of its angles, what may
it be called?

Name all the quadrilaterals.

Name three trapeziums.

Why is Fig. 5 a trapezium?

What is a trapezium?

Name two trapezoids.

Why is Fig. 6 a trapezoid?

Name its parallel sides.

What is a trapezoid?

Name six parallelograms.

Why is Fig. 4 a parallelogram?

Name its two pairs of parallel sides.

What is a parallelogram?

What two names can you give to Fig. 4?

Why the first? Why the second?

What two names may be given to Fig. 7?

Why the first? Why the second?

What two to Fig. 6?

Why the first? Why the second?

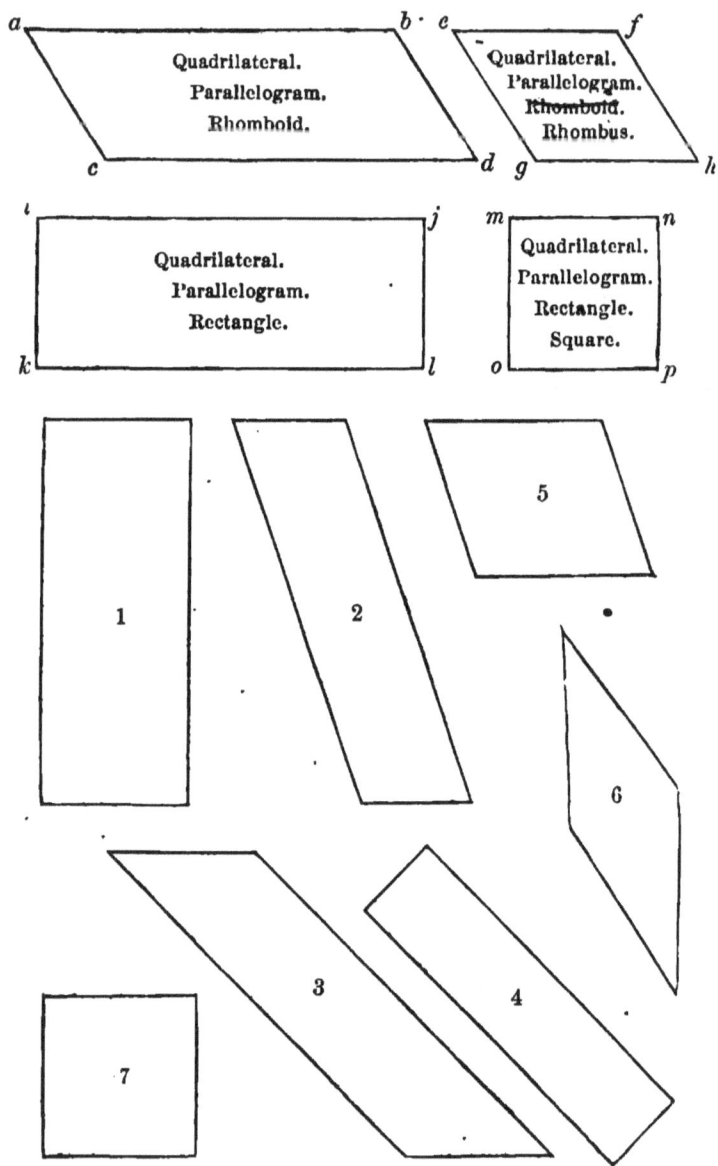

a b e f

Quadrilateral.
Parallelogram.
Rhomboid.

c d

Quadrilateral.
Parallelogram.
Rhomboid.
Rhombus.

g h

i j m n

Quadrilateral.
Parallelogram.
Rectangle.

k l o

Quadrilateral.
Parallelogram.
Rectangle.
Square.

p

1

2

5

6

3

4

7

Diagram 21.

LESSON TWENTY-SIXTH.

KINDS OF PARALLELOGRAMS.

RHOMBOIDS.

How many quadrilaterals in the diagram? (DIAGRAM 21.)

How many parallelograms?

Has the parallelogram *a d* any right angle?

It is called a "*rhomboid.*"

A rhomboid is a parallelogram which has no right angle.

Name five other rhomboids.

What three names may be given to Fig. 2?

Why is it a quadrilateral?

Why a parallelogram? Why a rhomboid?

RHOMBS.

Are the four sides of the rhomboid *a d* equal to each other?

Are the four sides of the rhomboid *e h* equal to each other?

If a triangle has its three sides equal to each other, what do you call it?

Then when a rhomboid has its sides equal to each other, what may it be called?

An equilateral rhomboid is called a rhombus.

A rhombus is an equilateral rhomboid.

See Note D, Appendix.

Name two other rhombuses, or rhombs.

What four names can you give to Fig. *e h?*

Why a quadrilateral? Why a parallelogram?
Why a rhomboid? Why a rhombus?

RECTANGLES.

Has the parallelogram *i l* any right angles?

How many?

It is called a "*rectangle.*"

A rectangle is a right-angled parallelogram.

Name four other rectangles.

What three names may be given to Fig. *i l?*

Why a quadrilateral? Why a parallelogram?
Why a rectangle?

SQUARES.

Has the rectangle *i l* its four sides equal?

Has the rectangle *m p* its four sides equal?

It is called a "square."

A square is an equilateral rectangle.

Name another "*square.*"

What four names may be given to Fig. *m p?*

Why a quadrilateral? Why a parallelogram?
Why a rectangle? Why a square?

LESSON TWENTY-SEVENTH.

REVIEW.

Name six rhomboids. (DIAGRAM 21.)

What three names may be given to Fig. 3?

Why a quadrilateral? Why a parallelogram? Why a rhomboid?

What is a quadrilateral? Parallelogram? Rhomboid?

Name three rhombs.

What four names may you give Fig. 5?

Why a quadrilateral? Why a parallelogram? Why a rhomboid? Why a rhomb?

What is a rhomboid? A rhomb?

Name five rectangles.

What three names may be given to Fig. 1?

Why a quadrilateral? Why a parallelogram? Why a rectangle?

What is a rectangle?

Name two squares?

By what four names may Fig. 7 be called?

Why by the first? By the second? By the third? By the fourth?

What is a square?

What is a rectangle?

What is a parallelogram?

What is a quadrilateral?

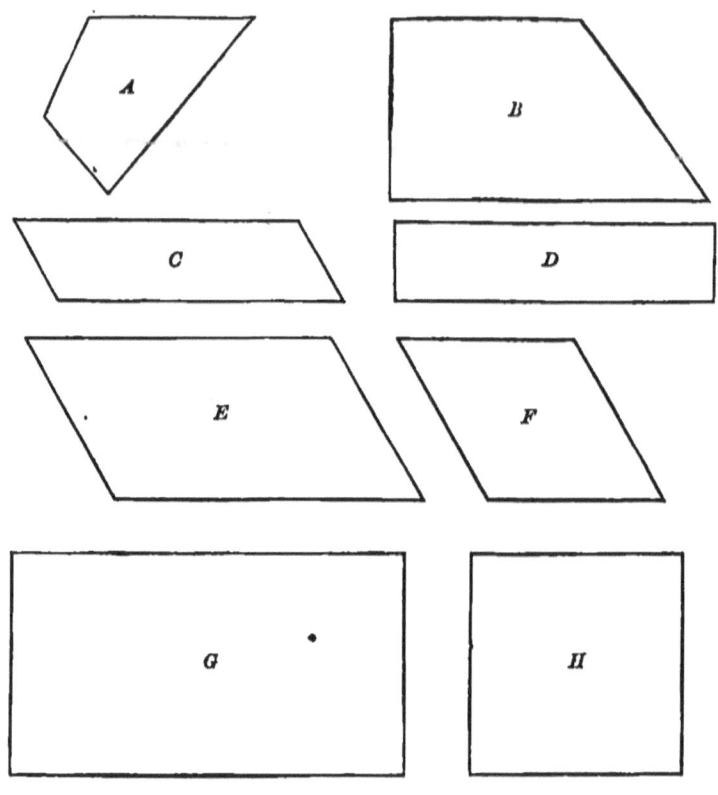

Diagram 22.

LESSON TWENTY-EIGHTH.

COMPARISON AND CONTRAST.

TRAPEZIUM AND TRAPEZOID.

In what respect are Figs. A and B alike?

On this account, what name may be given to each?

How does Fig. B differ from Fig. A?

What particular name may you give to Fig. B?

What one to Fig. A?

RHOMBOID AND RECTANGLE.

In what two respects are Figs. C and D alike?

On account of the number of their sides, what may each be called?

Because their opposite sides are parallel, what may each be called?

In what respect do they differ?

What particular name may be given to Fig. C?

What one to Fig. D?

What three names may you give to the figure with right angles?

What three to the one *without* right angles?

RHOMBOID AND RHOMBUS.

In what three things are Figs. E and F alike?

What three names may be given to each?

How do they differ from each other?

What particular name may you give to Fig. F?

What four names has Fig. F?

RECTANGLE AND SQUARE.

In what three things are Figs. G and H alike?

On account of the number of their sides, what may each be called?

Because their opposite sides are parallel, what may each be called?

Because they have right angles, what may they be called?

In what respect is Fig. H different from Fig. G?

On this account, what particular name may be applied to Fig. H?

What three names may be applied to Fig. G?

What *four* to Fig. H?

RHOMBUS AND SQUARE.

In what three things are Figs. F and H alike?

On account of the number of their sides, what name may be given to each?

Because their opposite sides are parallel, what name may be given to each?

Because both are parallelograms, and both have their sides equal, what name may be given to each?

What particular name has Fig. F?

What particular name has Fig. H?

What four names may be given to Fig. F?

What four to Fig. H?

LESSON TWENTY-NINTH.

REVIEW.

What two names may be given to Fig. A. (Dia-
gram 22.)
To Fig. B?
In what are they alike?
In what do they differ?
By what three names may Fig. C be called?
By what three names may Fig. D be called?
In what two things are they alike?
In what one thing do they differ?
What particular name has C? What one has D?
What three names may be applied to Fig. E?
What four to Fig. F?
What property has F that E has not?
What particular name has it on that account?
What three names has Fig. G?
What four has Fig. H?
What property has Fig. H that G has not?
What particular name has it in consequence?
What four names may you give to Fig. F?
What four to Fig. H?
What three names may be applied to either?
In what three things are they alike?
In what respect do they differ?
What particular name has Fig. F?
What particular name has Fig. H?

5

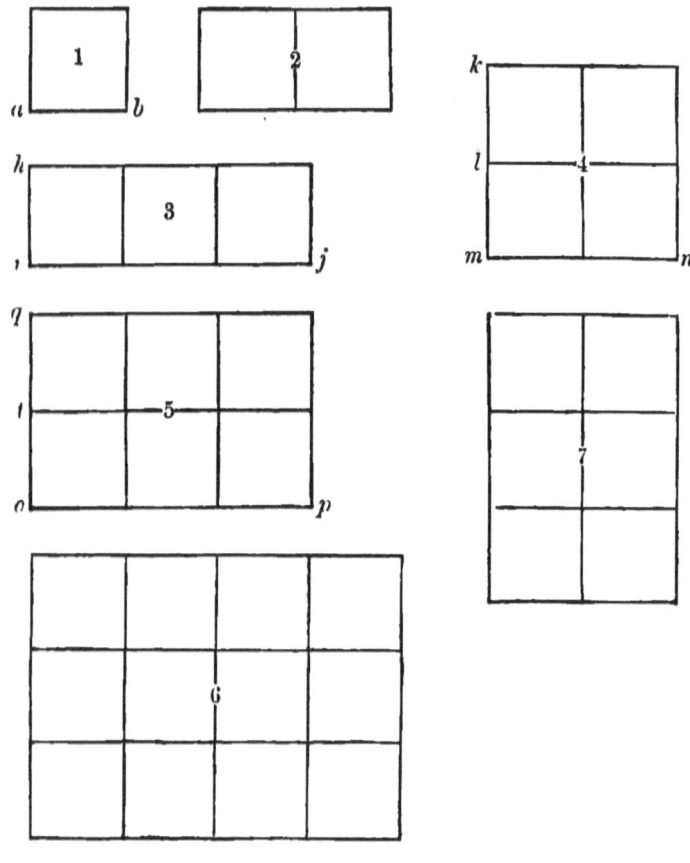

Diagram 23.

LESSON THIRTIETH.

MEASUREMENT OF SURFACES.

In Fig. 1 (DIAGRAM 23) call the line $a\,b$ a unit.
Rectangle 1 is how many units long?
How many high?
Because its sides are equal, what is it called?
Rectangle 2 is how many units long?

How many high or wide?

How many squares does it contain?

Rectangle 3 is how many units long?

How many wide?

How many squares does it contain?

If it were four units long and one wide, how many squares would it contain?

If it were five long and one wide? Six long? &c.

Rectangle 4 is how many units long?

How many wide?

How many squares does it contain?

How many squares in that part which is two units long, *m n*, and one unit wide, *m l?*

On account of the second unit in width, *l k*, how many times two squares are there?

If the width were one unit more, how many times two squares would there be?

Rectangle 5 is how many units long?

How many units wide or high?

How many squares does it contain?

How many squares in that part which is three units long, *o p*, and one unit wide, *o t?*

The second unit in width, *t q*, gives how many more squares? How many times three squares?

If another unit were added to the width, how many more squares would be made?

How many times three squares?

If it were four units wide, how many times three squares would there be?

Rectangle 6 is how many units long?

How many units high or wide ?

How many squares in that part which is four units long and one high ?

How many times four squares in that part which is four long and two high ?

How many times four squares when it is four long and three high ?

If another unit were added to the height, how many more squares would be added ?

How many times four squares would there be ?

If a rectangle were five units long and one unit wide, how many square units would it contain ?

If it were two units wide, how many times five square units would it contain ?

If it were three units wide ? Four ? &c.

If your ruler is ten inches long and only one inch wide, how many square inches are there in it ?

If it were two inches wide, how many times ten square inches would it contain ?

If your arithmetic-cover is seven inches long and five inches wide, how many square inches are there in it ?

If a wall of this room is twenty feet long, how many square feet are there in that part which is one foot high ? Two high ? Three high ? Four high ?

If the same wall is sixteen feet high, how many square feet in it ?

Fig. 5 has how many times three squares ?

Fig. 7 has how many times two squares?

Which has the greater number of squares?

What difference is there between two times three squares and three times two squares?

LESSON THIRTY-FIRST.

REVIEW.

Draw a rectangle of any width whose length is three times the width.

How many squares has it if the width be taken as the unit?

Make it twice as wide as before.

How many squares has it now?

What two numbers multiplied together will give the number of squares?

Make it three times as wide.

How many squares has it now?

What two numbers multiplied together will give the number of squares?

The cover of a geography is one foot long and one foot wide, how many square feet in it?

How many inches long is the same cover? How many wide?

How many square inches does it contain?

How many square inches are equal to one square foot?

A table is one yard long and one yard wide, how
  many square yards in it ?

How many feet long is the same table ? .

How many feet wide ?

How many square feet does it contain ?

One square yard equals how many square feet?

Draw a square whose side is a unit of any length.

Draw another whose side is two units of the same
  length.

The second square is how many times as large as
  the first one ?

How many squares in half the second square ?

Which is greater, two square inches, or two inches
  square ?

Two inches square is how many times two square
  inches ?

Draw a square whose side is three inches.

How many square inches does it contain ?

How many times as many squares as the square
  of one inch ?

How many square inches in the bottom row ?

How many in all ?

Which is greater, three inches square, or three
  square inches ?

Three inches square is how many times three
  square inches ?

PROBLEMS.

An equilateral triangle has each of its sides one inch long, what is its perimeter?

If each side were two inches long, what would be its perimeter?

An isosceles triangle has its two equal sides each three inches long, and its third side five inches long, what is its perimeter?

A right-angled isosceles triangle has its base five inches, and its hypothenuse seven inches long, what is its perimeter?

A square geography-cover is nine inches long on one side, how long all round?

How many square inches in it?

A slate is sixteen inches long and twelve wide, how many inches all round it?

A rectangle is five inches long and three wide, how long all round?

How many square inches in it?

A slate is one foot long and eight inches wide, what is its perimeter?

A room is twenty-four feet long and twenty-one feet wide, how many feet all round it?

How many square feet in the floor?

How many pieces of paper each a foot square would exactly cover it?

A yard of carpet is two feet wide, how many square feet in it?

Charles and Henry start from the same place, and
walk in opposite directions; Charles goes twenty
yards, and Henry fifteen, how many yards apart
are they?

If they start from opposite ends of a straight walk
twenty-five feet long, and walk towards each
other, how many feet will Charles have to walk
to meet Henry who has walked fifteen feet?

A lot is forty rods long and thirty wide, how long
must the fence be?

What length of fence will divide it into four
equal parts?

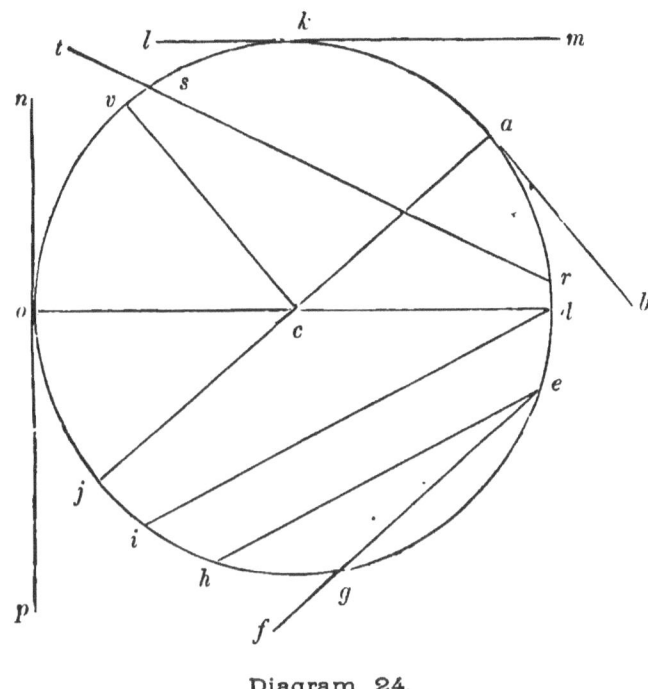

Diagram 24.

LESSON THIRTY-SECOND.

THE CIRCLE AND ITS LINES.

If the straight line *c a* were a string made fast at
c, with a sharp pencil-point at the other end *a*,
and the pencil-point were moved towards *d*, what
line would be drawn?

What kind of a line would it be?

If the pencil-point continued to move in the same
direction until it returned to the starting-point
a, what curved line would be drawn, naming it
by all the points in it which are marked?

The plane figure bounded by this curve is called a
"*circle*."

What point is at the centre of this figure?

*A circle is a plane figure bounded by a curved line,
all points of which are equally distant from the
centre.*

The curved line is called a "*circumference*."

*The circumference of a circle is the curve which
bounds it.*

Name a straight line that joins two points in the
circumference.

It is called a "*chord*."

*A chord is a straight line that joins two points of a
circumference.*

Read six chords in the diagram.

Which two of these chords pass through the centre?

They are called "*diameters*."

A diameter is a chord that passes through the centre.

Name a line that joins the centre with a point of
the circumference.

It is called a "*radius*." — (Plural, *radii*.)

*A radius is a straight line that joins the centre to a
point of the circumference.*

Read five radii.

Which is farther from the centre, the point *a* or
the point *d?*

Can the radius *c d* be greater than the radius *c a ?*
Or greater than *c v,* or *c o?*

Then all radii of the same circle are equal to each other.
What do we call the lines *o d, c d, c o?*
What part of the diameter *o d* is the radius *o c?*
Name a chord that is produced without the circle.
It is called a "*secant.*"

A secant is a chord produced.

Name two secants.
If the chord *d i* were made a secant, would it
    become longer or shorter?
In how many points does the straight line *l m*
    touch the circumference?
It is called a "*tangent.*"

A tangent is a straight line that touches a circumfer-
    ence in only one point.

Name three tangents.

LESSON THIRTY-THIRD.

REVIEW.

Read six chords. (DIAGRAM 24.)
Why is *i d* a chord?
What is a chord?
Name two diameters.
Why is *a j* a diameter?
What is a diameter?
Is every chord a diameter?

Is every diameter a chord?

Name five radii.

Why is *c a* a radius?

What is a radius?

A diameter is equal to how many radii?

Are all radii equal to each other?

Are all chords equal to each other?

Are all diameters equal to each other?.

Name two secants.

Why is either one a secant?

What is a secant?

Name three tangents.

Why is *a b* a tangent?

What is a tangent?

Is a tangent inside of a circle or outside of it?

Is a chord inside or outside of a circle?

Is a secant within or without a circle?

If the radius is three inches, how long is the diameter?

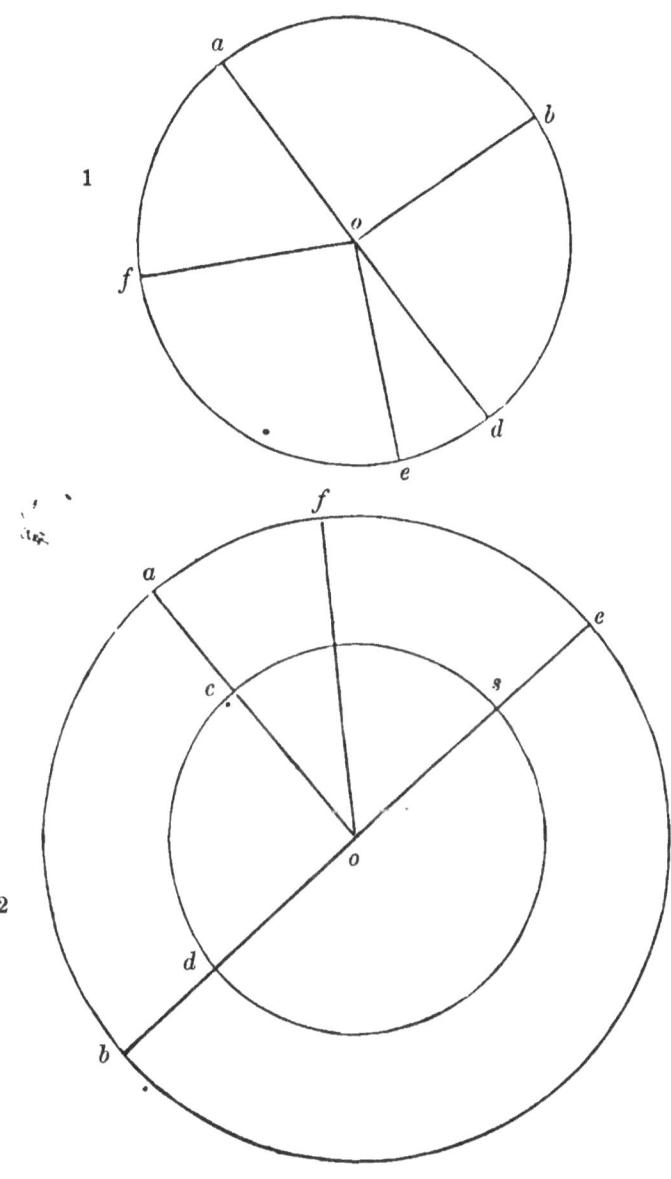

Diagram 25.

LESSON THIRTY-FOURTH.

ARCS AND DEGREES.

What small part of the circumference of circle 1
  (DIAGRAM 25) is marked?

It is called an "*arc.*"

An arc is any part of a circumference.

Read five arcs that are marked.

Which is longer, the arc *e d*, or the arc *e f? b d*, or
  b d e? a b d, or *a b d e?*

Name an arc which is half of the circumference.

It is called a "*semi-circumference.*"

"Semi" means "half."

A semi-circumference is half of a circumference.

Read three arcs, each of which is one-fourth of the
  circumference.

If the whole circumference were divided into three
  hundred and sixty equal arcs, would each arc be
  large or small?

Each of these arcs would be called a "*degree.*"
  [Degrees are marked (°).]

*A degree of a circumference is a three hundred and
  sixtieth part of it.*

How many degrees in a semi-circumference?

How many degrees in one-fourth of a circumference?

If a fourth of a circumference were divided into
  three equal parts, how many degrees would
  there be in each part?

Into how many parts would each third of a quarter have to be again divided to make single degrees?

Is an arc of ninety-one degrees greater or less than one-fourth of a circumference?

Is an arc of a hundred and seventy-nine degrees greater or less than a semi-circumference?

Can there be more than three hundred and sixty degrees in a circumference?

If the circumference of circle 1 were divided into degrees, each degree would be so small an arc that it would look like a dot.

If a degree were divided into sixty equal parts, each part would be called a minute.

If a minute were divided into sixty equal parts, each part would be called a second.

How many degrees in the large circle of Fig. 2?

How many in the smaller one?

Has a large circle any more degrees than a small circle?

In the large circle how many degrees from a to b?

In the small circle how many from a to b?

Which is greater, an arc of ninety degrees of the large circle, or one of ninety degrees of the small one?

Which is greater, an arc of a degree of the large circle, or one of a degree of the small one?

The angle $a\,o\,b$ has its vertex at what part of the larger circle?

At what part of the smaller circle?

On how many degrees of the larger circle does the angle stand?

On how many degrees of the smaller circle does it
stand ?

Then it is said to be an angle of 90°.

If the angle *a o f* is an angle of 30°, how many
degrees must there be in the arc *a f ?*

If the arc *f e* is an arc of 60°, what is the size of
the angle *f o e ?*

An angle of 10° stands upon an arc of how many
degrees ? Of 8° ? Of 1° ?

The angle *a o b* is what kind of an angle ?

Upon how many degrees does it stand ?

Then a right angle is an angle of how many
degrees ?

If an angle stand upon less than 90°, what kind of
an angle is it ?

If an angle stand upon more than 90°, what kind
of an angle is it ?

Can an angle have as many degrees as a hundred
and eighty ?

LESSON THIRTY-FIFTH.

REVIEW.

Read nine arcs whose ends are marked. (DIAGRAM 26.)

Read three arcs each of which is one-fourth of a circumference.

Read two arcs each of which is one-half of a circumference.

Why is $e\,g$ an arc?

What is an arc?

How many degrees in the arc $f\,h\,?$ In $e\,h\,?$

If the arc $f\,h$ were divided into three equal parts, how many degrees would there be in each?

How many degrees in a circumference?

In a semi-circumference?

How many more degrees in a large circumference than in a small one?

If the arc $i\,f$ is 40°, what is the size of the angle $f\,o\,i\,?$

If the angle $f\,o\,g$ is an angle of 130°, what is the size of the arc $f\,i\,h\,g\,?$

How many degrees in each of the adjacent angles $f\,o\,h,\,h\,o\,e\,?$

When two adjacent angles are equal to each other, what is each called?

How many degrees in a right angle?

6

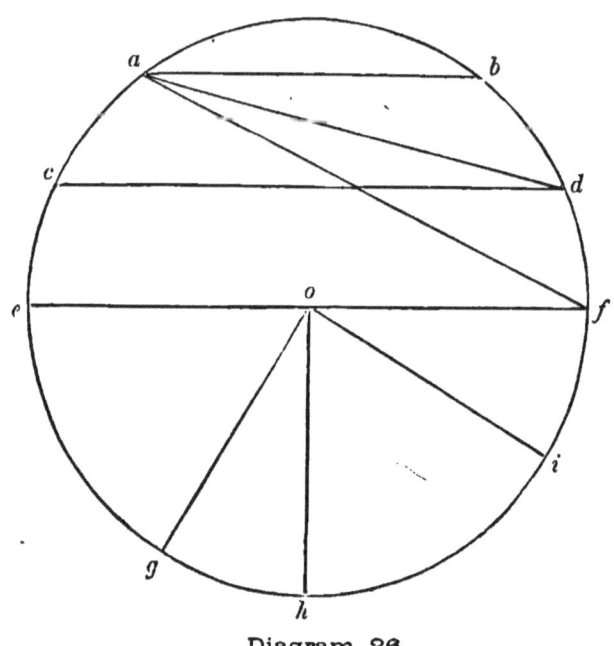

Diagram 26.

LESSON THIRTY-SIXTH.

PARTS OF THE CIRCLE.

The part of the circle bounded by the chord *a b*
and the arc *a b* is called a segment.

Read three segments, each less than half a circle,
thus, — the segment bounded by the chord *a d*
and the arc *a b d.*

*A segment is a part of a circle bounded by an arc
and a chord.*

Read two segments that are each half a circle.
What is the chord called ?
What is the arc called ?

A segment bounded by a diameter and a semi-circumference is a "*semicircle.*"

A semicircle is half a circle.

Read four segments each larger than a semicircle.

The part of the circle between the two radii *o f*, *o i*, and the arc *f i*, is called a "*sector.*"

Read four sectors each less than one-fourth of a circle.*

A sector is a part of a circle bounded by two radii and an arc.

What part of the whole circle is the sector *f o h?*
It is called a "*quadrant.*"

A quadrant is a sector which is one-fourth of a circle.

Read a sector which is greater than a quadrant.

If the chord *e f* be regarded a diameter, what do you call the semicircle below it?

If it be regarded as two radii, what is the semi-circle called?

Then a semicircle is both a segment and a sector.

---

* Thus, a sector bounded by the two radii *o g*, *o h*, and the arc *g h*.

LESSON THIRTY-SEVENTH.

•

REVIEW.

Name ten segments.   (DIAGRAM 26.)

What is a segment ?

Of the segments named, which are less than a
  semicircle ?

Which are greater ?

Which two are semicircles ?

Which two are on the chord af ?

Name nine sectors.

Why is $g\ o\ i$ a sector ?

What is a sector ?

Which four of the sectors named are each less than
  a quadrant ?

Which three are quadrants ?

Which two are greater than a quadrant ?

What part of the circle is both a segment and a
  sector ?

How many quadrants in a circle ?

How many semicircles ?

PART SECOND.

AXIOMS AND THEOREMS.

AXIOMS ILLUSTRATED.

AXIOM 1.

The triangle A is equal
to the triangle C.
The triangle B is also
equal to the triangle C.
What do you think of the two
triangles A and B?   Why?

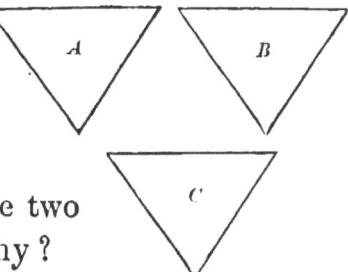

*If two things are separately equal to the same thing,
they are equal to each other.*

AXIOM 2.

The square A is equal to
the square B.
To the rectangle C add the
square A, and we have an
L pointing in what direc-
tion?
To the same rectangle C
add the square B, and we
have an L pointing in what direction?

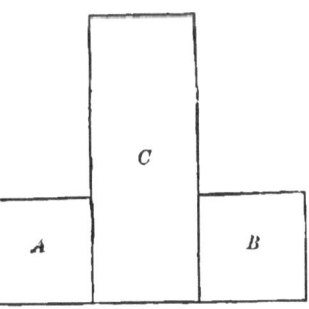

Which is larger, the L pointing to the left, or that
pointing to the right?

To what same thing did you add two equals?

What two equals did you add to it?

What was the first sum?

The second?

What do you think of the two sums?

*If equals be added to the same thing, the sums will
be equal.*

AXIOM 3.

The square A is equal to
the square B.

From the inverted T take
away the square A, and
we have an L pointing in
what direction?

From the same Fig. T take
away the square B, and
we have an L pointing in what direction?

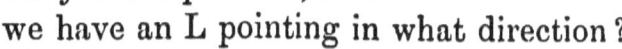

Which is larger, the L pointing to the right, or
that pointing to the left?

What two equal things did we take away from the
same thing?

From what same thing did we take them away?

What did we find true of the two remainders?

*If equals be taken from the same thing, the remain-
ders will be equal.*

AXIOM 4.

The rectangle 1 2 is equal to the rectangle 1 3.

From the rectangle 1 2 take away the square A, and what rectangle remains?

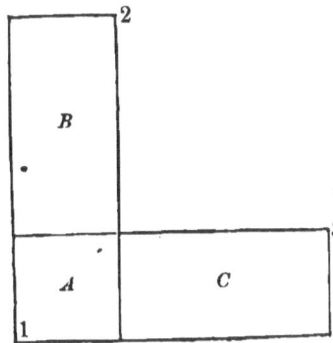

From the rectangle 1 3 take away the same square A, and what rectangle remains?

Which is greater, the rectangle B, or the rectangle C ?

What same thing did we take away from equals ?

From what did we first take it ?

What remained ?

From what did we next take it ?

What remained ?

What did we find true of the two remainders ?

If the same thing be taken from equals, the remainders will be equal.

AXIOM 5.

If equals be added to equals, the sums will be equal.

AXIOM 6.

If equals be subtracted from equals, the remainders will be equal.

AXIOM 7.

If the halves of two things are equal, the wholes will be equal.

Axiom 8.

Every whole is equal to the sum of all its parts.

Axiom 9.

From one point to another only one straight line can be drawn.

Axiom 10.

A straight line is the shortest distance between two points.

Axiom 11.

If two things coincide throughout their whole extent, they are equal.

THEOREMS ILLUSTRATED.

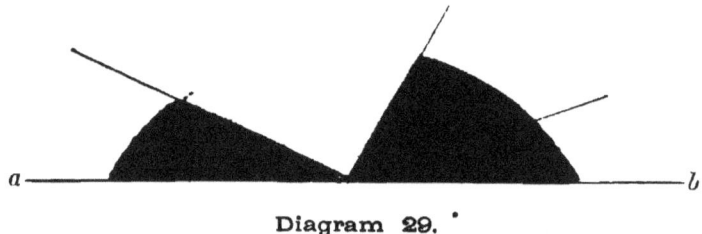

Diagram 29.

DEVELOPMENT LESSON.

Do the angles Blue, Red, take up all the space on the line *a b*?

Do the angles Blue, Yellow, Red, take up all the space on the line?

Do the angles Blue, Yellow, Green, Red, take up all the space on the line ?

Is there room between any two of the angles to put in another angle ?

Then are not the angles Blue, Yellow, Green, Red, equal to all the space on the line *a b ?*

NOTE. — The word *space*, as here used, means *angular space ;* and it is indispensable that the teacher impress this fact upon the learner.

By means of former lessons, the pupil has learned positively, that an angle is the difference between the directions of two lines; and, impliedly, that the included space has nothing to do with the size of the angle. There cannot, therefore, be much danger that the pupil will imbibe any erroneous notion from this style of expression, which is very much more simple than to say that the difference of direction of two given lines is equal to the difference of direction of two other given lines, which style will be used somewhat later in these lessons.

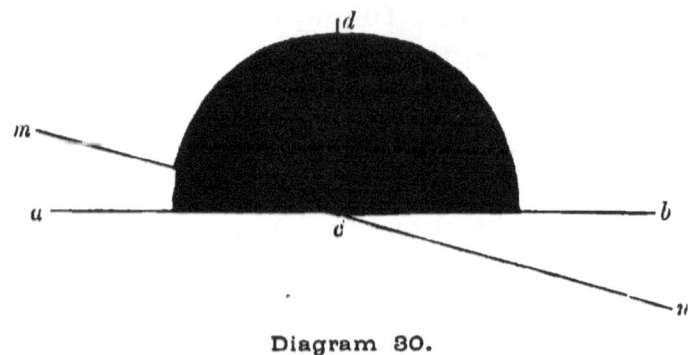

Diagram 80.

PROPOSITION I.   THEOREM.

DEVELOPMENT LESSON.

Are the adjacent angles Green, Red, equal to all
the angular space on the line $a\,b$?

Place a paper square corner or right angle on the
line $a\,b$ at the *left* of $c\,d$ with its vertex at c.

It will cover all the angle Green and part of the
angle Red up to the line $c\,d$.

Now place another square corner on the line $a\,b$ to
. the *right* of the line $c\,d$, and with its vertex at
the point c.

It will cover the remaining part of the angle Red,
and two edges of the square corners will meet
along the line $c\,d$.

Are the two right angles equal to all the angular
space on the line $a\,b$?

Then if the two adjacent angles Green, Red, are
equal to all the angular space on the line $a\,b$,
and the two right angles are also equal to the

same space, what do you infer concerning the *adjacent angles* and the *two right angles?*

What axiom do you apply when you say that the *adjacent* angles are equal to the *two right angles?*

To what *same thing* did you find two things separately equal?

What did you first see equal to it?

What did you next see equal to it?

Then what did you *find* true?

If the angle Red were smaller, and the angle Green larger, would the adjacent angles still be equal to two right angles?

Then, —

Any two adjacent angles are equal to two right angles.

If we draw the straight line *c d* where the edges of the square corners come together, what kind of angles will *a c d, d c b,* be?

See now if you can understand the following demonstration : —

DEMONSTRATION.

We wish to prove that

Any two adjacent angles are equal to two right angles.

Let the two straight lines *a b, m n,* intersect each other in the point *c.* (DIAGRAM 30.)

Then will any two adjacent angles, as Green, Red, be equal to two right angles?

For, from the point *c*, draw the straight line *c d* so
  as to make the angles *a c d*, *d c b*, right angles.

The adjacent angles Green, Red, are equal to all
· the angular space on the line *a b*.

The right angles *a c d*, *d c b*, are also equal to all
  the angular space on the line *a b*.

Therefore the adjacent angles Green, Red, are
  equal to two right angles.

TEST QUESTIONS.

To what same thing did you find two things equal?
What did you first see equal to it?
What did you next see equal to it?
Then what new thing did you find true?
What axiom did you make use of?

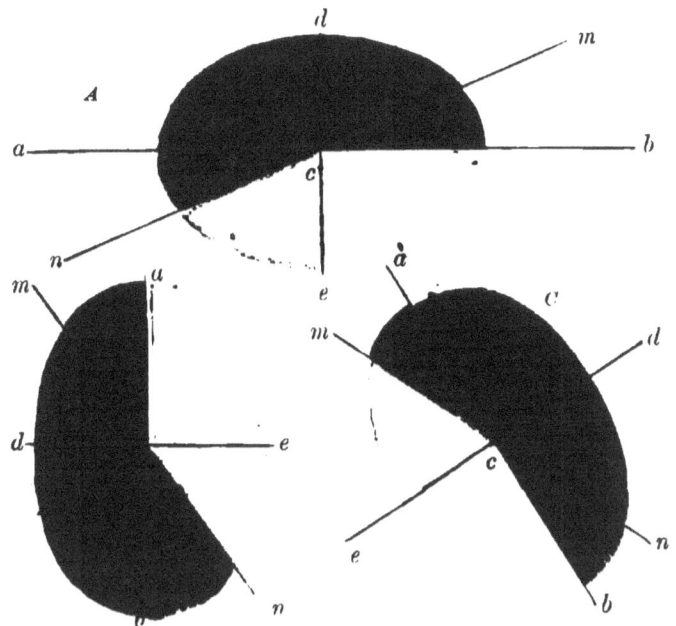

Diagram 31.

TEST LESSON.

By means of Fig. A, —

1. Prove that the adjacent angles Green, Red, are equal to two right angles.
2. Prove that the adjacent angles Blue, Yellow, are equal to two right angles.

By means of Fig. B, —

3. Prove that the adjacent angles Green, Red, are equal to two right angles.
4. Prove that the adjacent angles Yellow, Blue, are equal to two right angles.

By means of Fig. C, —

5. Prove that the adjacent angles Red, Blue, are equal to two right angles.

6. Prove that the adjacent angles Green, Yellow, are equal to two right angles.

7. Give the preceding demonstrations again, but name the angles by their letters instead of by their colors.

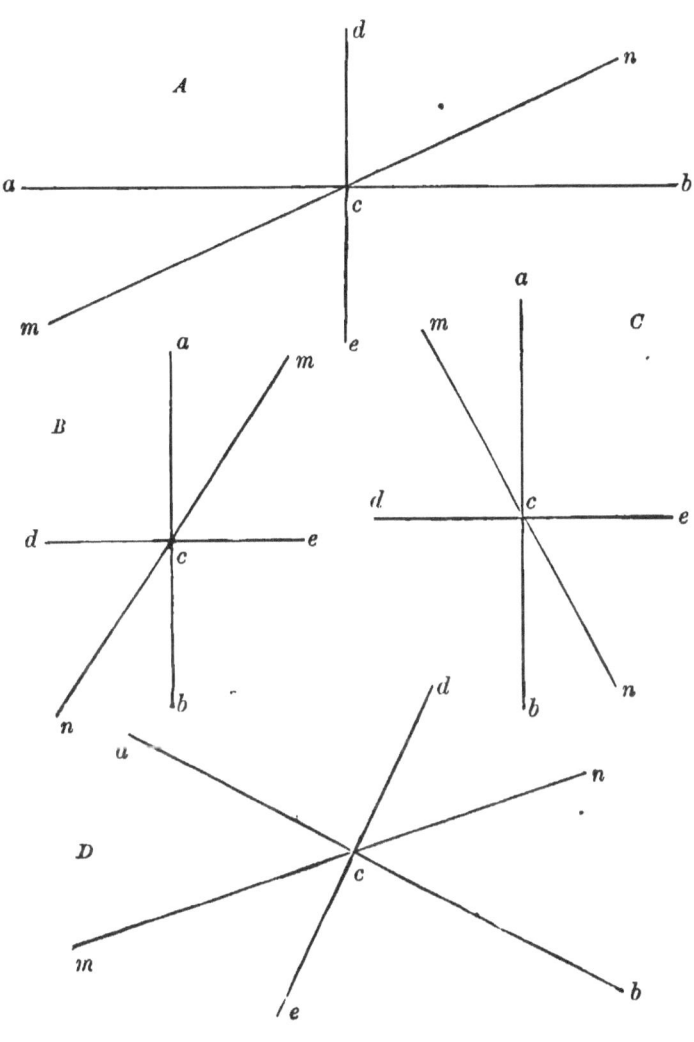

Diagram 32.

TEST LESSON.

By means of Fig. A prove, —

1. That the adjacent angles *a c m*, *m c b*, are equal to two right angles.
2. That the adjacent angles *a c n*, *n c b*, are equal to two right angles.

By means of Fig. B prove, —

3. That the adjacent angles *a c n*, *n c b*, are equal to two right angles.
4. That the adjacent angles *a c m*, *m c b*, are equal to two right angles.

By means of Fig. C prove, —

5. That the adjacent angles *a c m*, *m c b*, are equal to two right angles.
6. That the adjacent angles *a c n*, *n c b*, are equal to two right angles.

By means of Fig. D prove, —

7. That the adjacent angles *a c n*, *n c b*, are equal to two right angles.
8. That the adjacent angles *b c m*, *m c a*, are equal to two right angles.

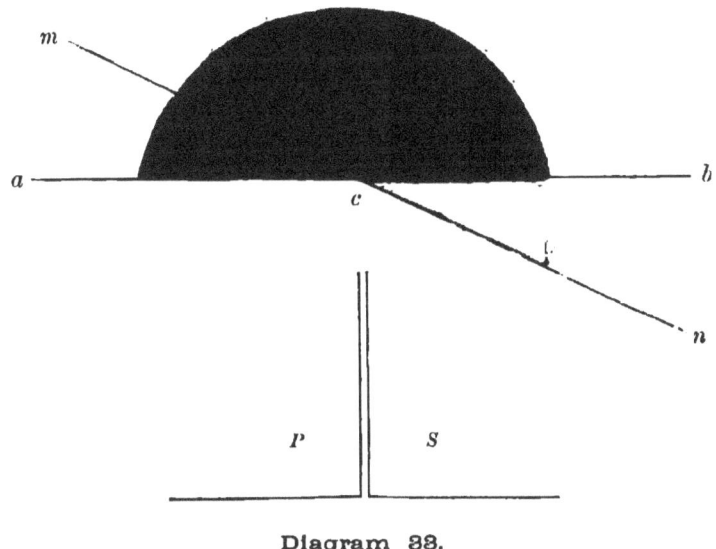

Diagram 88.

PROPOSITION II. THEOREM.

DEVELOPMENT LESSON.

What kind of angles are P and S?

How do the adjacent angles Yellow, Blue, compare with the right angles P, S?

How do the adjacent angles Blue, Red, compare with the two right angles?

Then if the adjacent angles Yellow, Blue, are equal to two right angles, and the adjacent angles Blue, Red, are also equal to two right angles, what do you think of the two pairs of adjacent angles, Yellow, Blue, and Blue, Red?

If, from the adjacent angles Yellow, Blue, we take away the angle Blue, what remains?

If, from the adjacent angles Blue, Red, we take away the same angle Blue, what remains?

Then, since the same angle Blue has been taken from equal pairs of adjacent angles, what do you think of the two remainders, Yellow, Red?

Suppose the lines *a b* and *m n* were so drawn that the angles Yellow, Red, were larger or smaller, would they still be equal to each other?

Then, —

All vertical angles are equal to each other.

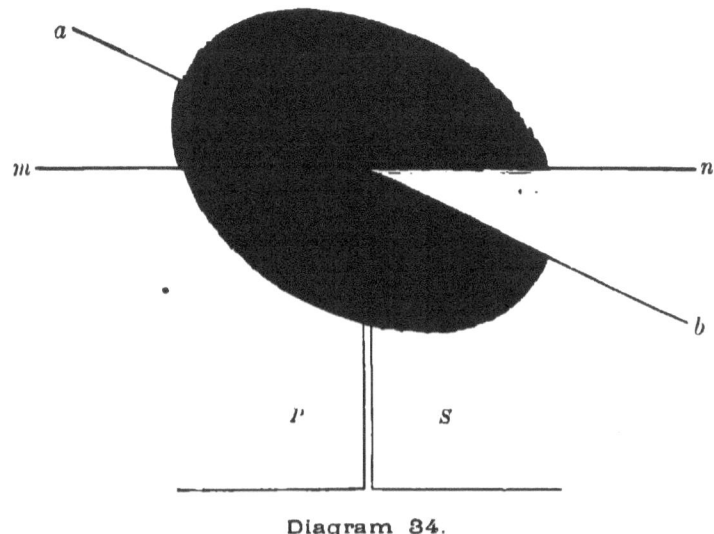

Diagram 34.

DEMONSTRATION.

We wish to prove that

All vertical angles are equal to each other.

Let the straight lines *a b, m n*, intersect each other at the point *c*, then will any two vertical angles, as Yellow, Red, be equal to each other.

For the adjacent angles Yellow, Blue, are equal to two right angles.*

The adjacent angles Blue, Red, are also equal to two right angles.

Therefore the adjacent angles Yellow, Blue, are equal to the adjacent angles Blue, Red.

If, from the adjacent angles Yellow, Blue, we take away the angle Blue, we shall have left the angle Yellow.

---

* When this comparison is made, let the pupil look at the right angles P and S.

If, from the adjacent angles Blue, Red, we take away the same angle Blue, we shall have left the angle Red.

Therefore the vertical angles Yellow, Red, are equal to each other.

TEST QUESTIONS.

When you say that the adjacent angles Yellow, Blue, are equal to two right angles, do you know it because you *see* it, or because you have *proved* it?

How do you know that the adjacent angles Blue, Red, are equal to two right angles?

When you say the adjacent angles Yellow, Blue, are equal to the adjacent angles Blue, Red, what axiom do you use?

What same thing do you take away from equals?

From what equals do you take it away?

When you take the angle Blue from the adjacent angles Yellow, Blue, what is the remainder?

When you take the same angle Blue from the adjacent angles Blue, Red, what is the remainder?

What do you find true of the two remainders?

What axiom do you use?

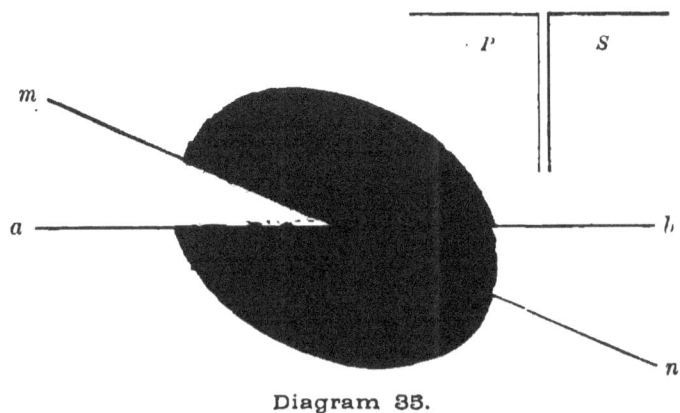

Diagram 85.

OTHER METHODS OF DEMONSTRATION.

The adjacent angles Yellow, Green, are equal to what?

The adjacent angles Green, Red, are equal to what?

Then what do you know of the two pairs of adjacent angles Yellow, Green, and Green, Red?

From the adjacent angles Yellow, Green, take away the angle Green.   What remains?

From the adjacent angles Green, Red, take the same angle Green.   What remains?

What do you know of the two remainders?

Why?

What axiom do you use?

In the last lesson, when you proved the vertical angles Yellow, Red, equal to each other, you made use of the angle Blue; now prove the same two angles equal by means of the angle Green.

The adjacent angles Blue, Red, are equal to what?

The adjacent angles Red, Green, are equal to what?

Then what do you know of the two pairs of adjacent angles Blue, Red, and Red, Green?

From the adjacent angles Blue, Red, take away the angle Red.   What remains?

From the adjacent angles Red, Green, take away the same angle Red.   What remains?

Then what do you know of the two remainders, Blue, Green?

Now apply the preceding demonstration to the vertical angles Blue, Green.

Prove the vertical angles Blue, Green, equal to each other by means of the angle Yellow.

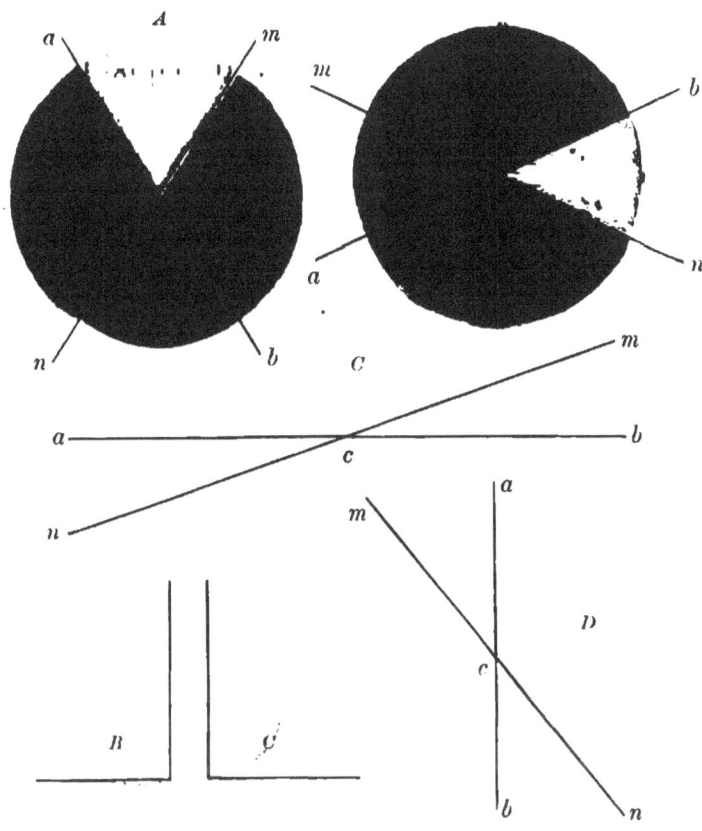

Diagram 86.

TEST LESSON.

By means of Fig. A, —

1. Prove that the vertical angles Yellow, Red, are equal to each other, using the angle Green.
2. Prove the same thing, using the angle Blue.
3. Prove that the vertical angles Blue, Green, are equal to each other, using the angle Yellow.
4. Prove the same thing, using the angle Red.

By means of Fig. B, —

> 5. Prove the vertical angles Yellow, Red, equal to each other, using the angle Green.
> 6. Prove the same thing, using the angle Blue.
> 7. Prove the vertical angles Green, Blue, equal by means of the angle Red.
> 8. Prove the same thing by means of the angle Yellow.

Go through the preceding eight demonstrations again, calling the angles by their letters instead of by their colors.

By means of Fig. C, prove that

> 9. $a\,c\,n$ equals $m\,c\,b$, by means of $a\,c\,m$.
> 10. $a\,c\,n$ equals $m\,c\,b$, by means of $b\,c\,n$.
> 11. $a\,c\,m$ equals $n\,c\,b$, by means of $a\,c\,n$.
> 12. $a\,c\,m$ equals $n\,c\,b$, by means of $m\,c\,b$.

By means of Fig. D, prove that

> 13. $m\,c\,a$ equals $b\,c\,n$, by means of $a\,c\,n$.
> 14. $m\,c\,a$ equals $b\,c\,n$, by means of $m\,c\,b$.
> 15. $m\,c\,b$ equals $a\,c\,n$, by means of $m\,c\,a$.
> 16. $m\,c\,b$ equals $a\,c\,n$, by means of $b\,c\,n$.

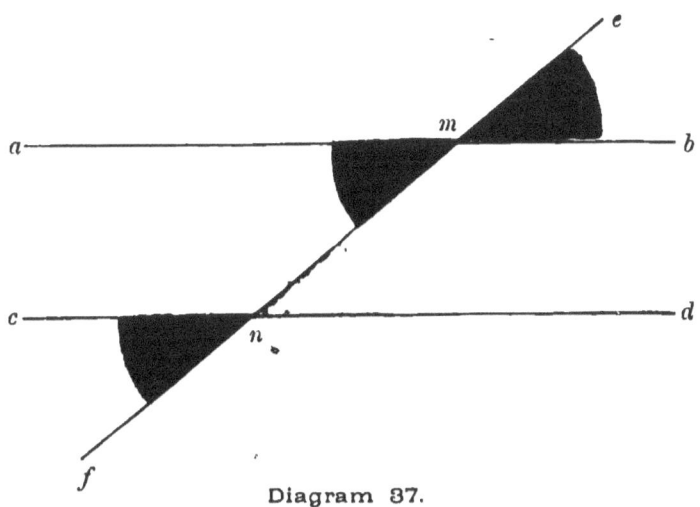

Diagram 37.

PROPOSITION III.   THEOREM.

DEVELOPMENT LESSON.

In the above diagram, the lines *a b, c d*, are parallel, and are intersected by the line *e f* at the
points *m* and *n*.

The angle Red measures the difference of direction between the line *m b* and what other line ?

The angle Yellow measures the difference of direction between the line *n d* and what other line ?

Then, as the lines *m b* and *n d* are parallel, must
there not be the same difference of direction
between them and the line *e f?*

Then can there be any difference between the
angles which measure those equal directions ?

Then what do you think of the opposite exterior
and interior angles Red, Yellow ?

DEMONSTRATION.

We wish to prove that

Opposite exterior and interior angles are equal to each other.

Let the straight line $e f$ intersect the two parallel straight lines $a b$, $c d$, at the points m and n .

Then will any two opposite exterior and interior angles, as Red, Yellow, be equal to each other.

For the angle Red measures the difference of direction of the lines $m b$ and $e f$.

And the angle Yellow measures the difference of direction of the lines $n d$ and $e f$.

But because the lines $m b$, $n d$, are parallel, these differences are equal.

Therefore the angles which measure them are equal; that is,

The opposite exterior and interior angles Red, Yellow, are equal to each other.

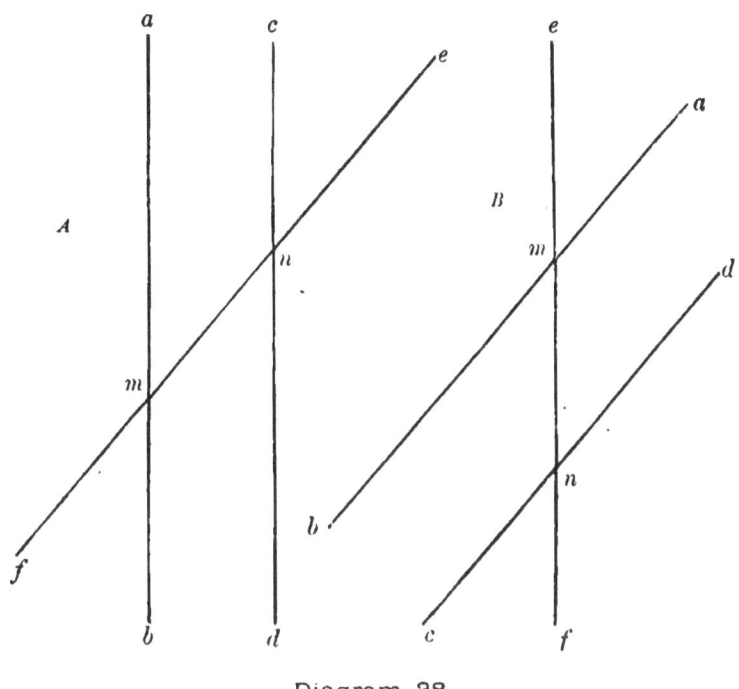

Diagram 38.

TEST LESSON.

By means of Fig. A, —

1. Prove that the opposite exterior and interior angles Green, Blue, are equal to each other.
2. Prove that the opposite exterior and interior angles Red, Yellow, are equal to each other.
3. Prove the opposite exterior and interior angles *c n e, a m n,* equal.
4. Prove the opposite exterior and interior angles *e n d, n m b,* equal.

By means of Fig. B, —

5. Prove the opposite exterior and interior angles *e m a, m n d,* equal.

6. Prove the opposite exterior and interior angles *a m n, d n f,* equal.

7. Prove the opposite exterior and interior angles *e m b, m n c,* equal.

8. Prove the opposite exterior and interior angles *b m n, c n f,* equal.

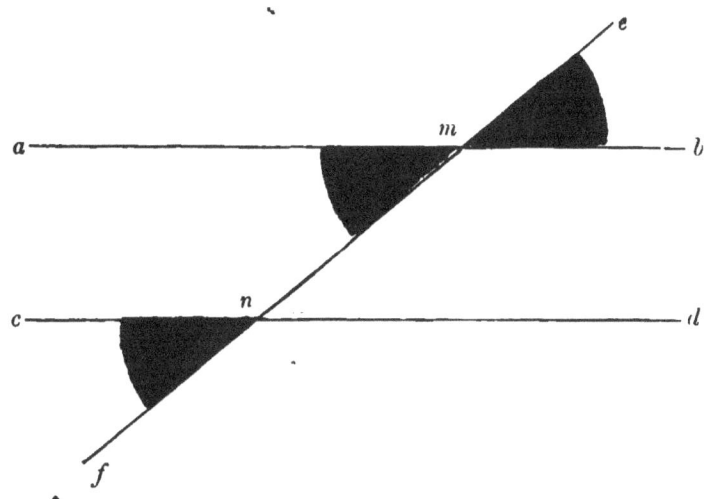

Diagram 39.

PROPOSITION IV. THEOREM.

DEVELOPMENT LESSON.

What do you know of the opposite exterior and
  interior angles Red, Yellow?

What do you know of the vertical angles Red,
  Green?

Then if the interior alternate angles Green, Yellow, are separately equal to the angle Red, what
  new fact do you know?

What axiom do you employ?

To what same thing did you find two things equal?

What two things did you find equal to it?

DEMONSTRATION.

We wish to prove that

Any two interior alternate angles are equal to each other.

Let the straight line ef intersect the two parallel straight lines $a\ b$, $c\ d$, in the points m and n.

Then will any two interior alternate angles, as
· Green, Yellow, be equal to each other.

For the opposite exterior and interior angles Red, Yellow, are equal.

The vertical angles Red, Green, are also equal.

Then because the interior alternate angles Green, Yellow, are separately equal to the angle Red, they are equal to each other.

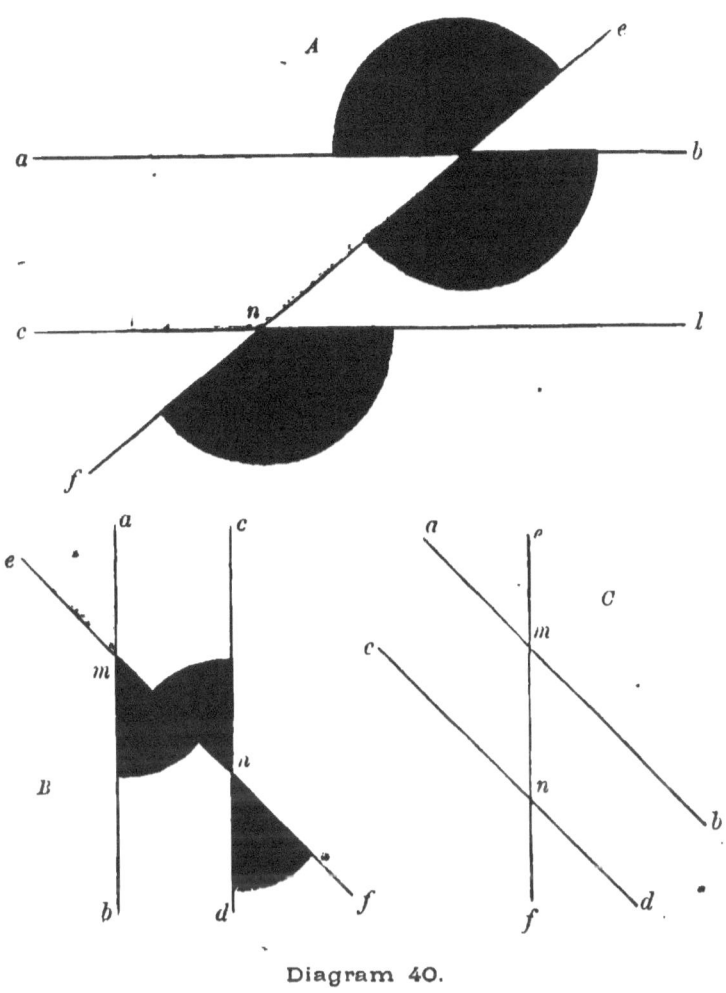

Diagram 40.

TEST LESSON.

What do you know of the vertical angles Green, Red, in Fig. A?

What do you know of the opposite exterior and interior angles Red, Yellow?

Then if the interior alternate angles Green, Yel-

low, are separately equal to the angle Red, what
do you infer?

By means of Fig. A, —

1. Prove that the interior alternate angles Green,
Yellow, are equal, using the angle Red.
2. Prove the same angles equal, using the angle
Blue.
3. Go through the same demonstrations again,
calling the angles by their letters instead of by
their colors.

By means of Fig. B, —

4. Prove the interior alternate angles Red, Blue,
equal, using the angle Yellow.
5. Prove the same angles equal, using the angle
Green.
6. Go through the same two demonstrations again,
naming the angles by their letters instead of by
their colors.

By means of Fig. C, —

7. Prove the interior alternate angles $c\,n\,m$, $n\,m\,b$,
equal, using the angle $f\,n\,d$.
8. Prove the same, using the angle $a\,m\,e$.
9. Prove the interior alternate angles $a\,m\,n$, $m\,n\,d$,
equal, using the angle $e\,m\,b$.
10. Prove the same, using the angle $c\,n\,f$.

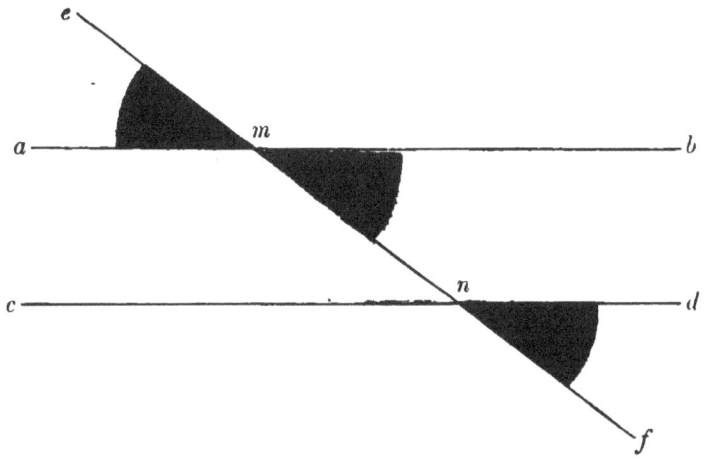

Diagram 41.

PROPOSITION V. THEOREM.

DEVELOPMENT LESSON.

What do you know of the opposite exterior and interior angles Red, Yellow?

What do you know of the vertical angles Yellow, Green?

Then if the exterior alternate angles Red, Green, are separately equal to the angle Yellow, what new thing do you know to be true?

What axiom do you employ?

To what same thing did you know two things to be equal?

What two things did you know to be equal to it?

Then what new thing did you *find* to be true?

8

DEMONSTRATION.

We wish to prove that

Any two exterior alternate angles are equal to each other.

Let the straight line *e f* intersect the two parallel straight lines *a b*, *c d*, at the points *m* and *n*.

Then will any two exterior alternate angles, as Red, Green, be equal.

For the opposite exterior and interior angles Red, Yellow, are equal to each other.

And the vertical angles Yellow, Green, are also equal to each other.

Then because the exterior alternate angles Red, Green, are separately equal to the angle Yellow, they are equal to each other.

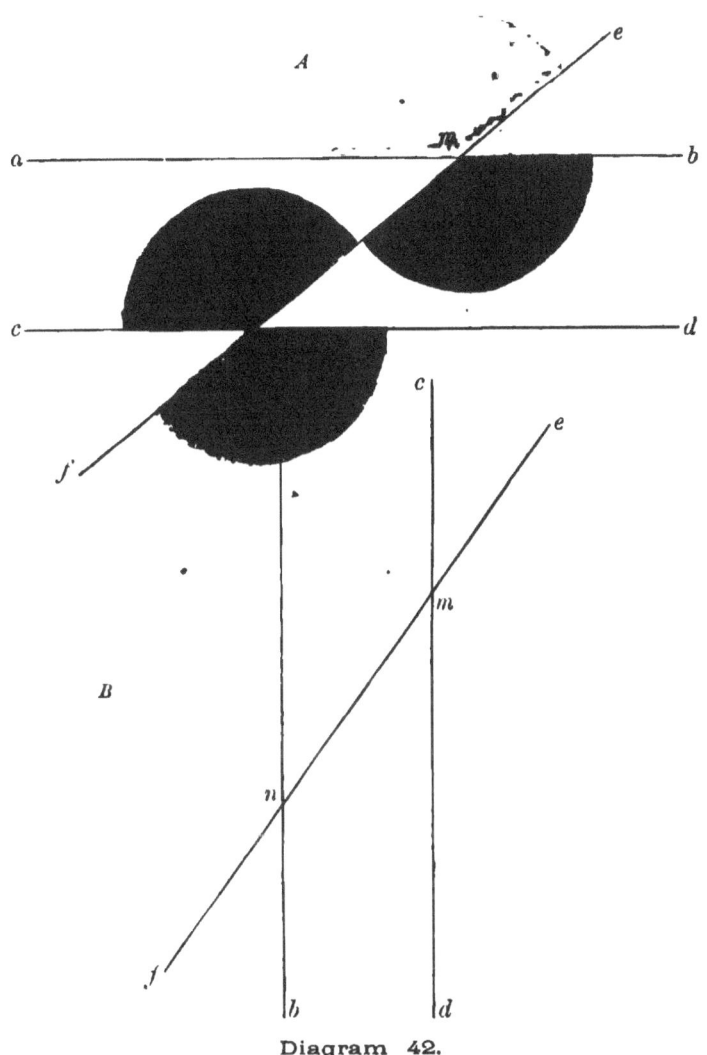

Diagram 42.

TEST LESSON.

What do you know of the opposite exterior and interior angles Yellow, Red ?

What do you know of the vertical angles Red, Blue ?

Then if the exterior alternate angles Yellow, Blue,
are separately equal to the angle Red, what do
you know of them?

By means of Fig. A, —

1. Prove that the exterior alternate angles Yellow,
Blue, are equal, using the angle Red.
2. Prove the same thing, using the angle Green.
3. Go through the same demonstrations, calling the
angles by their letters.
4. Prove the exterior alternate angles *e m b, c n f,*
equal, using the angle *a m n.*
5. Prove the same, using the angle *m n d.*

By means of Fig. B, —

6. Prove that the exterior alternate angles *c m e,
f n b,* are equal, using the angle *n m d.*
7. Prove the same, using the angle *a n m.*
8. Prove the exterior alternate angles *e m d, a n f,*
equal, using the angle *c m n.*
9. Prove the same, using the angle *m n b.*

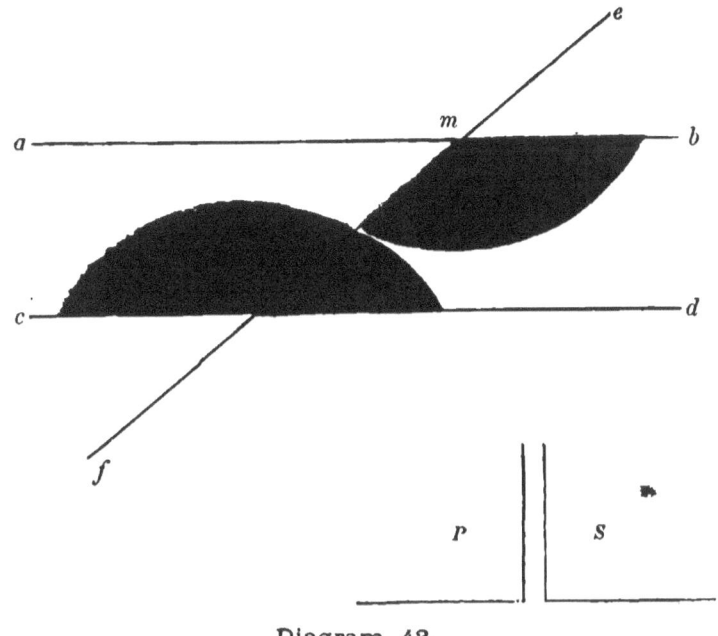

Diagram 48.

PROPOSITION VI. THEOREM.

DEVELOPMENT LESSON.

What do you know of the interior alternate angles Yellow, Red?

If to the angle Green you add the angle Yellow, what is the sum?

If to the same angle Green you add the equal angle Red, what is the sum?

Then, having added equals to the same thing, what do you think of the two sums, — the adjacent angles Green, Yellow, and the interior opposite angles Green, Red?

What do you know of the adjacent angles Green, Yellow, and the right angles P, S?

Then if the interior opposite angles Green, Red,
and the two right angles P, S, are separately
equal to the adjacent angles Green, Yellow,
what new thing do you know?

DEMONSTRATION.

We wish to prove that

*Any two interior opposite angles are equal to two
right angles.*

Let the straight line *e f* intersect the two parallel
straight lines *a b, c d*, in the points *m* and *n*.

Then will any two interior opposite angles be
equal to two right angles.

For the interior alternate angles Yellow, Red, are
equal.

If to the angle Green we add the angle Yellow,
we shall have the adjacent angles Green, Yellow.

If to the same angle Green we add the equal angle
Red, we shall have the interior opposite angles
Green, Red.

Then the adjacent angles Green, Yellow, are equal
to the interior opposite angles Green, Red.

·But the adjacent angles Green, Yellow, are equal
to two right angles.

Then because the interior opposite angles Green,
Red, and two right angles, are separately equal
to the two adjacent angles Green, Yellow, they
are equal to each other.

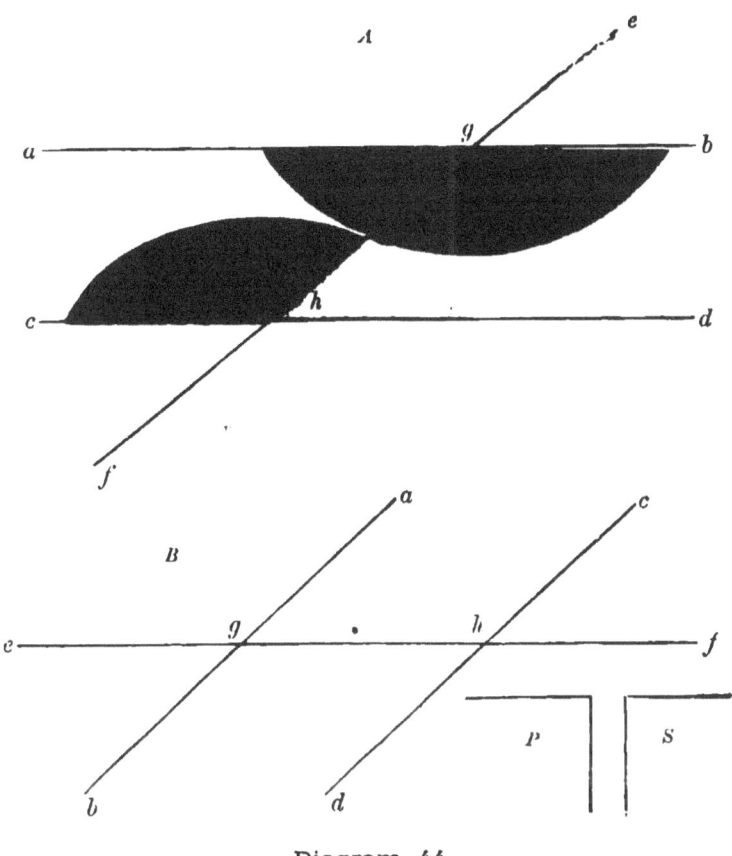

Diagram 44.

TEST LESSON.

By means of Fig. A, —

1. Prove the interior opposite angles Green, Yellow, equal to two right angles, using the angle Red.
2. Prove the same, using the angle Blue.
3. Prove the same, using the angle *e g b*.
4. Prove the same, using the angle

5. Go through the same demonstrations again, naming the angles by their letters instead of by their colors.

6. Prove the interior opposite angles Red, Blue, equal to two right angles, using the angle Yellow.

7. Prove the same, using the angle Green.

8. Prove the same, using the angle $e\,g\,a$.

9. Prove the same, using the angle $c\,h\,f$.

10. Go through the same demonstrations again, calling the angles by their letters instead of by their colors.

By means of Fig. B, —

11. Prove the interior opposite angles $a\,g\,h$, $g\,h\,c$, equal to two right angles, using the angle $g\,h\,d$.

12. Prove the same, using the angle $c\,h\,f$.

13. Prove the same, using the angle $a\,g\,e$.

14. Prove the interior opposite angles $b\,g\,h$, $g\,h\,d$, equal to two right angles, using the angle $a\,g\,h$.

15. Prove the same, using the angle $e\,g\,b$.

16. Prove the same, using the angle $f\,h\,d$.

Compare the angles Yellow, Green, each with its exterior opposite angle, and see if you can prove that the exterior opposite angles $e\,g\,b$, $f\,h\,d$, are also equal to two right angles.

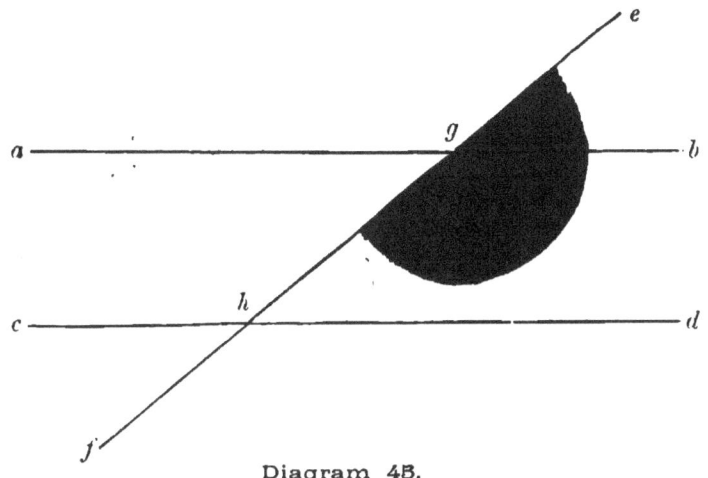

Diagram 48.

PROPOSITION VII. THEOREM.

DEVELOPMENT LESSON.

Suppose we do not know whether the lines $a\,b, c\,d,$
   are parallel, or not;

But, by measuring, we find that the interior angles
   Blue, Yellow, on the same side of the secant*
   line $e\,f,$ are equal to two right angles:

The adjacent angles Blue, Red, are equal to what?

Then, if the interior angles Blue, Yellow, are equal
   to two right angles,

And the adjacent angles Blue, Red, are also equal
   to two right angles,

What do you infer?

From the interior angles Blue, Yellow, take away
   the angle Blue: what remains?

From the adjacent angles Blue, Red, take away the
   same angle Blue: what remains?

* "Secant" means "cutting."

What do you know of the two remainders?

The angle Red measures the direction of the line
   g b from what line?

The equal angle Yellow measures the direction of
   the line *h d* from what line?

Then if the lines *g b, h d,* have the same direction
   from the line *e f,* what do you call them?

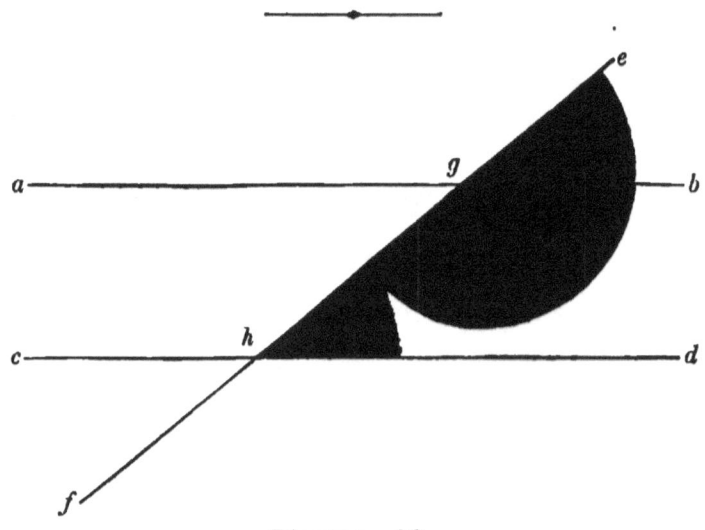

Diagram 46.

DEMONSTRATION.

We wish to prove, that,

*If a straight line intersects two other straight lines so
that two interior angles on the same side of the
intersecting line are equal to two right angles, the
two lines are parallel.*

Let the straight line *e f* intersect the two straight
   lines *a b, c d,* in the points *g* and *h,* so that the
   angles Red, Blue, are equal to two right angles.

Then will the lines *a b, c d,* be parallel.

For the angles Red, Blue, are supposed equal to two right angles.

The adjacent angles Red, Green, are known to be also equal to two right angles.

Then the interior angles Red, Blue, are equal to the adjacent angles Red, Green.

If from the interior angles Red, Blue, we take away the angle Red, we have left the angle Blue.

If from the adjacent angles Red, Green, we take the same angle Red, we shall have left the angle Green.

Then the angle Blue is equal to the angle Green.

But the angle Blue measures the direction of the line *h d* from the line *e f.*

And the angle Green measures the direction of the line *g b* from the line *e f.*

Then the lines *g b, h d,* have the same direction, and are parallel.

TEST LESSON.

1. Prove the same without the colors.
2. Prove the same, using the angle *f h d.*
3. Prove the same, supposing the angles *a g h, g h c,* equal to two right angles, and using the angle *a g e.*
4. Prove the same, using the angle *c h f.*

See Note E, Appendix.

PROPOSITION VIII. THEOREM.

The following demonstration is very easy. Read it once, and see if you can go through it without a second reading : —

DEMONSTRATION.

We wish to prove that

The sum of any two sides of a triangle is greater than the third side.

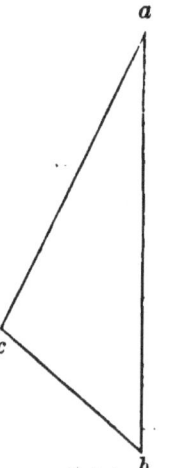

Let the figure *a b c* be a triangle, then will the sum of any two sides, as *a c, c b,* be greater than the third side *a b.*

For the straight line *a b* is the shortest distance between the two points *a* and *b*, and is therefore less than the broken line *a c b.*

PROPOSITION IX. PROBLEM.

The following solution is so easy that you will understand it at once : —

We wish

To construct an equilateral triangle on a given straight line.

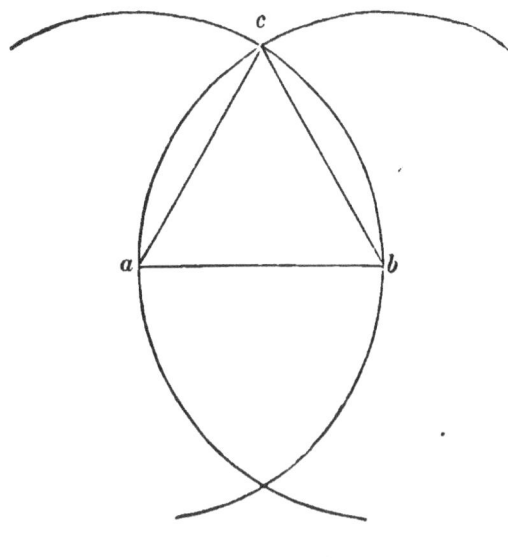

SOLUTION.

Let *a b* be the given line.

With the point *a* as a centre, and *a b* as a radius, draw the circumference of the circle, or a part of one.

With the point *b* as a centre, and the same radius *a b*, draw another circumference, or a part of one.

From the point *c*, in which the circumferences or arcs intersect, draw the straight lines *a c* and *b c*.

Now, because the lines *a b* and *a c* are radii of the same circle, they are equal.

And, because the lines *a b* and *b c* are radii of the same circle, they are also equal.

Then, because the two lines *a c*, *b c*, are separately equal to the line *a b*, they are equal to each other, and the triangle is equilateral.

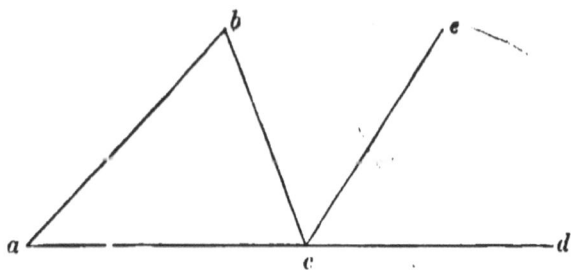

PROPOSITION X. THEOREM.

DEVELOPMENT LESSON.

Let the figure *a b c* be a triangle.

Produce the side *a c* to *d*.

We have now another angle, *b c d*, and we wish to find out if it is equal to any of the angles of the triangle.

From the point *c* draw the line *c e* parallel to *a b*.

Because the straight line *a d* intersects the two parallels *a b, c e*, the angle *a* is equal to what other angle?

Because the straight line *b c* intersects the two parallels *a b, c e,* the angle *b* is equal to what other angle?

Then the angles *a* and *b* are equal to what two angles?

How does the angle *b c d* compare with the angles *b c e, e c d?*

Then, if the angles *a* and *b*, on the one hand, and the angle *b c d*, on the other, are separately equal to the angles *b c e, e c d,*

What have you found out?

What axiom have you just employed?

To what same thing have you found two other things equal?

What two things did you find equal to it?

DEMONSTRATION.

We wish to prove, that,

If any side of a triangle be produced, the new angle formed will be equal to the sum of the angles that are not adjacent to it.

Let $a\,b\,c$ be a triangle.

Produce the side $a\,c$ to d; then will the new angle $b\,c\,d$ be equal to the sum of the angles a and b.

For from the point c draw $c\,e$ parallel to $a\,b$.

Then, because the straight line $a\,d$ intersects the two parallels $a\,b$, $c\,e$, in the points a and c,

The opposite exterior and interior angles a and $e\,c\,d$ are equal to each other.

And because the straight line $b\,c$ intersects the same parallels in the points b and c,

The interior alternate angles b and $b\,c\,e$ are equal.

Then the angles a and b of the triangle are equal to the angles $b\,c\,e$ and $e\,c\,d$.

But the new angle $b\,c\,d$ is equal to the angles $b\,c\,e$, $e\,c\,d$.

Then because the new angle $b\,c\,d$, and the angles a and b are separately $=$ to the angles $b\,c\,e$, $e\,c\,d$, they are equal to each other.

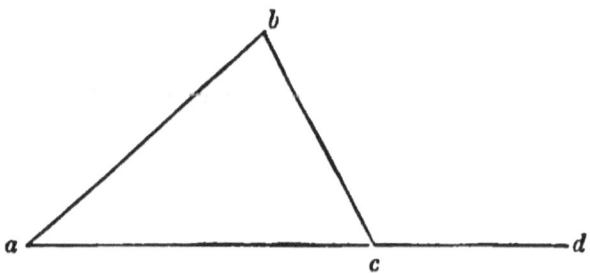

PROPOSITION XI.   THEOREM.

DEVELOPMENT LESSON.

Let the figure *a b c* be a triangle.

Produce the side *a c* to *d*.

By the last theorem, the angle *b c d* is equal to what angles of the triangle?

What angle must we add to these angles to make up the three angles of the triangle?

If we add the same angle to the angle *b c d*, what adjacent angles do we get?

Then the three angles of the triangle, *a*, *b*, and *c*, are equal to what two angles?

But the adjacent angles *a c b* and *b c d* are equal to what?

Then, because the three angles of the triangle, *a*, *b*, and *c*, and two right angles, are separately equal to the two adjacent angles *c* and *b c d*,

What new thing have you found out?

DEMONSTRATION.

We wish to prove that

The three angles of any triangle are equal to two right angles.

Let the figure *a b c* be a triangle; then will the sum of the angles *a, b,* and *c,* be equal to two right angles.

For, produce the side *a c* to *d,*

The new angle *b c d* is equal to the sum of the angles *a* and *b.*

If to the angles *a* and *b* we add the angle *c,* we shall have the three angles of the triangle.

If to the angle *b c d* we add the same angle *c,* we shall have the adjacent angles *c* and *b c d.*

Then the three angles of the triangle *a, b, c,* are equal to the adjacent angles *c* and *b c d.*

But the adjacent angles *c* and *b c d* are equal to two right angles.

Then, because the three angles of the triangle are equal to the adjacent angles *c* and *b c d,* they are equal to two right angles.

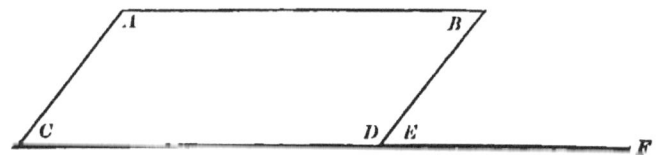

PROPOSITION XII. THEOREM.

DEVELOPMENT LESSON.

Let the Fig. A B C D be a parallelogram.

Produce the side C D to F.

Because the straight line B D intersects the parallels A B and C F, the angle B is equal to what other angle?

Because the straight line C F intersects the parallels A C and B D, the angle C is equal to what other angle?

Then what follows from this?

To what angle did you find two others equal?

What two angles did you find equal to it?

What axiom do you think of?

See if you can go through the demonstration without reading it even once.

DEMONSTRATION.

We wish to prove that

The opposite angles of a parallelogram are equal to each other.

Let the Fig. A B C D be a parallelogram.

Then will any two opposite angles, as B and C, be equal to each other.

For produce the line C D to F.

Because the straight line B D meets the two parallels A B and C F,

The interior alternate angles B and E are equal to each other.

Because the straight line C F meets the two parallels B D and A C,

The opposite exterior and interior angles C and E are equal to each other.

Then, because the angles B and C are separately equal to the angle E, they are equal to each other.

———

1. Prove the same by producing the line A B towards the left.
2. Prove the same by producing the line B D downwards.
3. Prove the angles A and D equal to each other by producing the line C D towards the left.
4. Prove the same by producing the line D B upwards.
5. See if you can prove the same by drawing a diagonal through the points A and D.

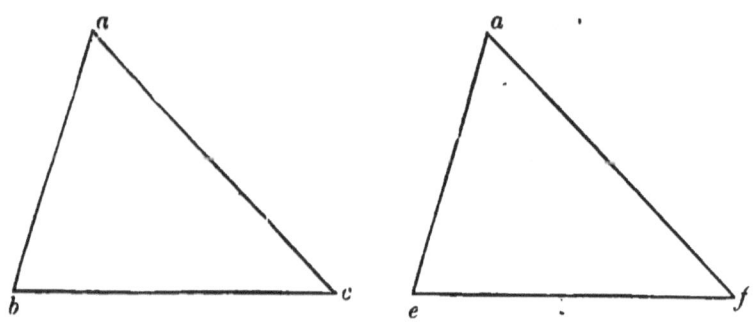

PROPOSITION XIII. THEOREM.

DEVELOPMENT LESSON.

In these two triangles we have tried to make the side *a b* of the one equal to the side *d e* of the other; the side *a c* of the one equal to the side *d f* of the other; and the included angle *b a c* of the one equal to the included angle *e d f* of the other.

We now wish to find out if the third side *b c* of the one is equal to the third side *e f* of the other, and if the two remaining angles *b* and *c* of the one are equal to the two remaining angles *e* and *f* of the other.

Suppose we were to cut the triangle *d e f* out of the page, and place it upon the triangle *a b c*, so that the line *d e* should fall upon the line *a b*, and the point *d* upon the point *a*.

As the line *d e* is equal to the line *a b*, upon what point will the point *e* fall?

If the angle *e d f* were less than the angle *b a c*,

would the line $d\,f$ fall within or without the triangle ?

If the angle $e\,d\,f$ were greater than the angle $b\,a\,c$, where would the line $d\,f$ fall ?

Since the angle a is equal to d, where, then, must the line $d\,f$ fall ?

As the line $d\,f$ is equal to the line $a\,c$, upon what point will the point f fall ?

Then, if the point e falls upon the point b, and the point f upon the point c, where will the line $e\,f$ fall ?

Now, because the three sides of the triangle $d\,e\,f$ exactly fall upon the three sides of the triangle $a\,b\,c$, we say *the two magnitudes coincide throughout their whole extent*, and are therefore equal.

What three parts of the triangle $a\,b\,c$ did we suppose to be equal to three corresponding parts of the triangle $d\,e\,f$ before we placed one upon the other.

What line of the one do we *find* equal to a line in the other ?

What two angles of the one do we *find* equal to two angles in the other ?

What do you think of the areas of the triangles ?

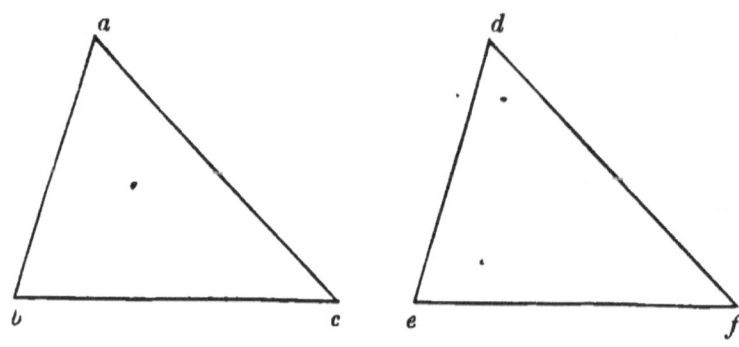

DEMONSTRATION.

We wish to prove, that,

If two triangles have two sides, and the included angle of the one equal to two sides and the included angle of the other, each to each, the two triangles are equal in all respects.

Let the triangles *a b c* and *d e f* have the side *a b* of the one equal to the side *d e* of the other; the side *a c* of the one equal to the side *d f* of the other; and the included angle *b a c* of the one equal to the included angle *e d f* of the other, each to each; then will the two triangles be equal in all their parts.

For, place the triangle *d e f* upon the triangle *a b c*, so that the line *d e* shall fall upon the line *a b*, with the point *d* upon the point *a*.

Because the line *d e* is equal to the line *a b*, the point *e* will fall upon the point *b*.

Because the angle *e d f* is equal to the angle *b a c*, the line *d f* will fall upon the line *a c*.

Because the line *d f* is equal to the line *a c*, the point *f* will fall upon the point *c*.

Then, because the point e is on the point b, and the
point f on the point c, the line ef will coincide
with the line $b\,c$, and the two triangles will be
found equal in all their parts;

That is, the angle e is found to be equal to the
angle b, the angle f to the angle c, the line ef
to the line $b\,c$, and the area of the triangle
$a\,b\,c$ to the area of the triangle $d\,e\,f$.

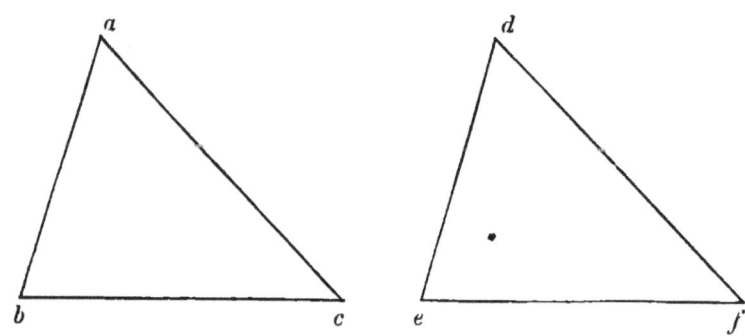

PROPOSITION XIV.   THEOREM.

DEVELOPMENT LESSON.

In these two triangles we have tried to make the
angle b of the one equal to the angle e of the
other; the angle c of the one equal to the angle
f of the other; and the included side $b\ c$ of
the one equal to the included side ef of the
other.

We now wish to find out if the remaining angle
a of the one is equal to the remaining angle d
of the other, and if the two remaining sides $a\ b$
and $a\ c$ of the one are equal to the two remaining sides $d\ e$ and df of the other.

Suppose we were to cut the triangle $d\ ef$ out of
the page and place it upon the triangle $a\ b\ c$, so
that the line ef shall fall upon the line $b\ c$, with
the point e upon the point b.

Because the line ef is equal to the line $b\ c$, upon
what point will the point f fall?

Because the angle e is equal to the angle b, where
will the line $e\ d$ fall?

Because the angle f is equal to the angle c, where will the line $d f$ fall?

Then, if the line $d e$ falls upon the line $a b$, and the line $d f$ upon the line $a c$, where will the point d fall?

Now because the three sides of the triangle $d e f$ exactly fall upon the three sides of the triangle $a b c$, we say *the two magnitudes coincide throughout their whole extent, and are therefore equal.*

Suppose the angle e were greater than the angle b, would the line $e d$ fall within or without the triangle?

If it were less, where would the line fall?

Why does the line $d e$ fall exactly upon the line $a b$?

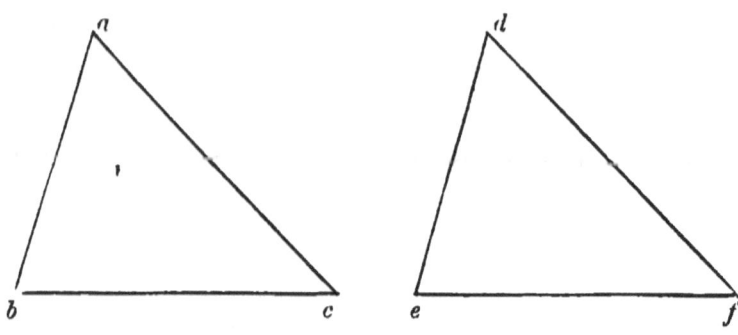

DEMONSTRATION.

We wish to prove that,

If two triangles have two angles, and the included side of the one equal to two angles and the included side of the other, each to each, the two triangles are equal to each other in all respects.

Let the triangles *a b c* and *d e f* have the angle *b* of the one equal to the angle *e* of the other; the angle *c* of the one equal to the angle *f* of the other; and the included side *b c* of the one equal to the included side *e f* of the other, each to each; then will the two triangles be equal in all their parts.

For place the triangle *d e f* upon the triangle *a b c*, so that the line *e f* shall fall upon the line *b c*, with the point *e* upon the point *b*.

Because the line *e f* is equal to the line *b c*, the point *f* will fall upon the point *c*.

Because the angle *e* is equal to the angle *b*, the line *e d* will fall upon the line *b a*, and the point *d* will be somewhere in the line *b a*.

Because the angle f is equal to the angle c, the line $f\,d$ will fall upon the line $c\,a$, and the point d will be somewhere in the line $c\,a$.

Then, because the point d is in the two lines, $b\,a$ and $c\,a$, it must be in their intersection, or upon the point a.

And, as the two triangles coincide throughout their whole extent, they are equal in all their parts.

That is, the angle a is found to be equal to the angle d; the side $b\,a$ to the side $e\,d$; the side $c\,a$ to the side $f\,d$; and the area of the triangle $a\,b\,c$ to the area of the triangle $d\,e\,f$.

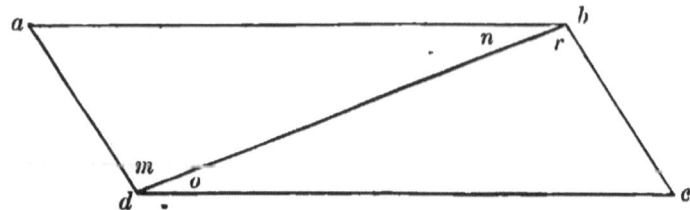

PROPOSITION XV.   THEOREM.

DEMONSTRATION.

We wish to prove that

The opposite sides of any parallelogram are equal.

Let the figure *a b c d* be a parallelogram; then will
the sides *a b* and *c d* be equal to each other;
likewise the sides *a d* and *b c*.

For, draw the diagonal *b d*.

Because the figure is a parallelogram, the sides
a b and *d c* are parallel, and the interior alternate angles *n* and *o* are equal.

Because the figure is a parallelogram, the interior
alternate angles *r* and *m* are equal.

Then the two triangles *a d b, b d c*, have two angles
and the included side of the one equal to two
angles and the included side of the other, each
to each, and are therefore equal;

And the side *a b* opposite the angle *m* is equal to
the side *c d* opposite the equal angle *r;*

And the side *a d* opposite the angle *n* is equal to
the side *b c* opposite the equal angle *o*.

TEST.

Prove the same by drawing a diagonal from *a* to *c*.

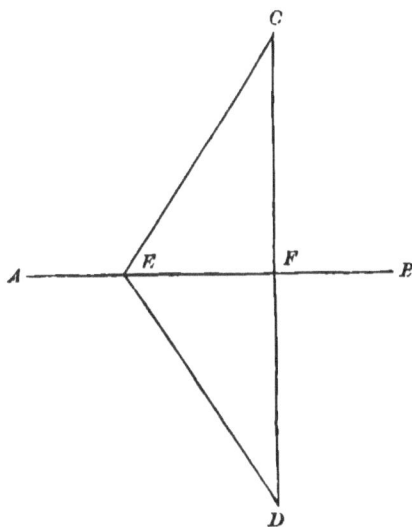

PROPOSITION XVI. THEOREM.

DEVELOPMENT LESSON.

Suppose A B to be a straight line, and C any point out of it.

From the point C draw a perpendicular C F to A B.

Let us see if this perpendicular is not shorter than any other line we can draw from the same point to the same line.

Draw any other line from C to A B as C E.

Now, as C E is any line whatever other than a perpendicular, if we find that the perpendicular C F is shorter than it we must conclude that it is the shortest line that can be drawn from C to A B.

Produce C F until F D is equal to C F, and then join E and D.

In the triangles E F C, E F D, what two sides were
drawn equal?

What line is a side to each?

How great an angle is C F E?

What is a right angle?

Then how do the angles C F E and E F D compare
with each other?

If the two triangles E F C, E F D, have the side C F
of the one equal to the side F D of the other,
the side E F common to both, and the included
angle E F C of the one equal to the included
angle E F D of the other, each to each, what do
you infer?

Then what third side of the one have you found
equal to a third side of the other?

C E is what part of the broken line C E D?

C F is what part of the line C D?

Which is shorter, the straight line C D, or the
broken line C E D?

Then how does the half of C D or C F compare with
the half of C E D or C E?

If C E is any line whatever other than a perpendicular, what may we now say of the perpendicular from the point C to the straight line A B?

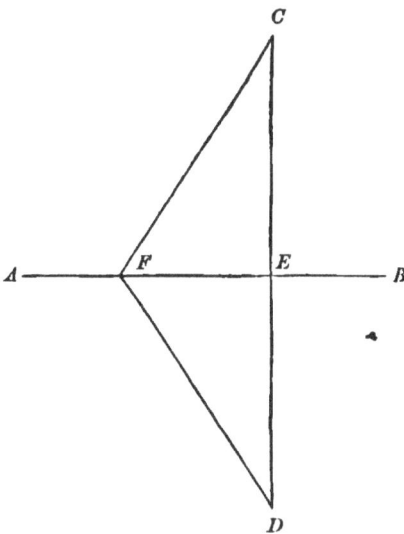

DEMONSTRATION.

We wish to prove that

A perpendicular is the shortest distance from a point to a straight line.

Let A B be a straight line, and C a point out of it; then will the perpendicular C E be the shortest line that can be drawn from the point to the line.

For draw any other line from C to A B, as C F.

Produce C E until E D equals C E, and join F D..

The two triangles F E C, F E D, have the side C E of the one equal to the side E D of the other, the side F E common, and the included angle F E.C of the one equal to the included angle F E D of the other, they are therefore equal, and the side C F equals the side F D.

But the straight line C D is the shortest distance between the two points C D; therefore it is shorter than the broken line C F D.

Then C E, the half of C D, is shorter than C F, the half C F D.

And, as C F is any line other than a perpendicular, the perpendicular C E is the shortest line that can be drawn from C to A B.

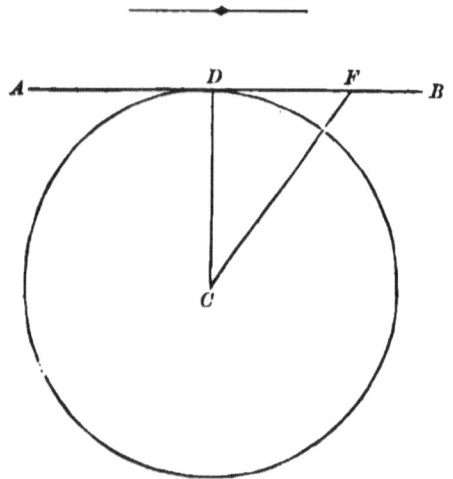

PROPOSITION XVII. THEOREM.

DEMONSTRATION.

We wish to prove that

A tangent to a circumference is perpendicular to a radius at the point of contact.

Let the straight line A B be tangent at the point D to the circumference of the circle whose centre is C.

Join the centre c with the point of contact D, the tangent will be perpendicular to the radius c D.

For draw any other line from the centre to the tangent, as c F.

As the point D is the only one in which the tangent touches the circumference, any other point, as F, must be without the circumference.

Then the line c F, reaching *beyond* the circumference, must be longer than the radius c D, which would reach only to it; therefore c D is shorter than any other line which can be drawn from the point c to the straight line A B; therefore it is perpendicular to it.

PROPOSITION XVIII.   THEOREM.

DEMONSTRATION.

We wish to prove, that,

In any isosceles triangle, the angles opposite the equal sides are equal.

Let the triangle A B C be isosceles, having the side A B equal to the side A C; then will the angle B, opposite the side A C, be equal to the angle C, opposite the equal side A B.

For draw the line A D so as to divide the angle A into two equal parts, and let it be long enough to divide the side B C at some point as D.

Now the two triangles A D B, A D C, have the side A B of the one equal to the side A C of the other, the side A D common to both, and the included angle B A D of the one equal to the included angle C A D of the other; therefore the two triangles are equal in all respects, and the angle B, opposite the side A C, is equal to the angle C, opposite the side A B.

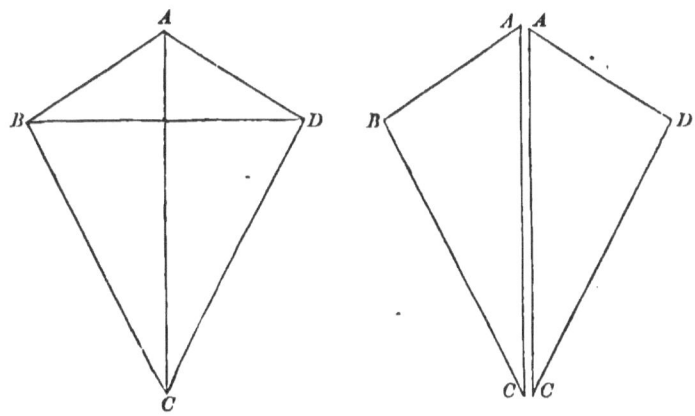

PROPOSITION XIX. THEOREM.

DEMONSTRATION.

We wish to prove that,

If two triangles have the three sides of the one equal to the three sides of the other, each to each, they are equal in all their parts.

Let the two triangles A B C, A D C, have the side A B of the one equal to the side A D of the other; the side B C of the one equal to the side D C of the other, and the third side likewise equal; then will the two triangles be equal in all their parts.

For place the two triangles together by their longest side, and join the opposite vertices B and D by a straight line.

Because the side A B is equal to the side A D, the triangle B A D is isosceles, and the angles A B D, A D B, opposite the equal sides are equal.

Because the side B C is equal to the side D C, the
triangle B C D is isosceles, and the angles C B D,
C D B, opposite the equal sides are equal.

If to the angle A B D we add the angle D B C, we
shall have the angle A B C.

And if to the equal of A B D, that is, A D B, we add
the equal of D B C, that is, B D C, we shall have
the angle A D C.

Therefore the angle A B C is equal to the angle
A D C.

Then the two triangles A B C, A D C, have two sides,
and the included angle of the one equal to two
sides and the included angle of the other, each
to each, and are equal in all their parts; that is,
the three angles of the one are equal to the
three angles of the other, and their areas are
equal.

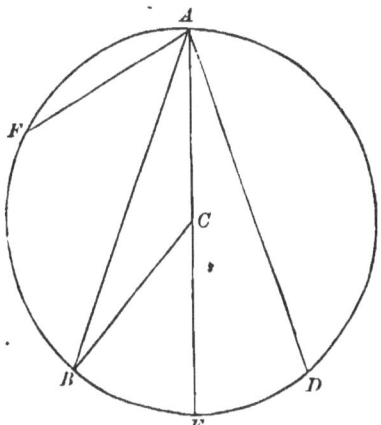

PROPOSITION XX. THEOREM.

DEMONSTRATION.

We wish to prove that

An angle at the circumference is measured by half the arc on which it stands.

Let B A D be an angle whose vertex is in the circumference of the circle whose centre is C; then will it be measured by half the arc B D.

For through the centre draw the diameter A E, and join the points C and B.

The exterior angle E C B is equal to the sum of the angles B and B A C.

Because the sides C A, C B, are radii of the circle, they are equal, the triangle is isosceles, the angles B and B A C opposite the equal sides are equal, and the angle B A C is half of both.

Then, because the angle B A C is half of B and B A C, it must be half of their equal E C B.

But E C B, being at the centre, is measured by B E; then half of it, or B A C, must be measured by half B E.

In like manner, it may be proved that the angle C A D is measured by half E D.

Then, because B A C is measured by half B E, and C A D by half E D, the whole angle B A D must be measured by half the whole arc B D.

SECOND CASE.

Suppose the angle were wholly on one side of the the centre, as F A B.

Draw the diameter A E and the radius B C as before.

Prove that the angle B A E is measured by half the arc B E.

Draw another radius from C to F, and prove that F A E is measured by half the arc F E.

Then, because the angle F A E is measured by half the arc F E, and the angle B A E is measured by half the arc B E,

The difference of the angles, or F A B, must be measured by half the difference of the arcs, or half of F B.

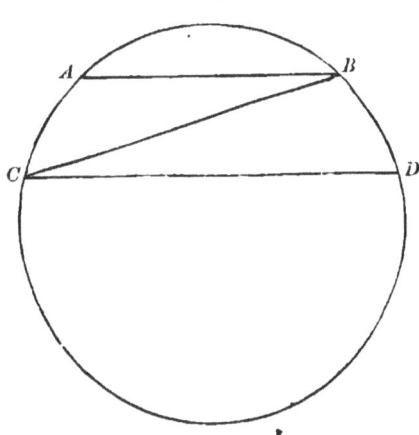

PROPOSITION XXI.   THEOREM.

DEMONSTRATION.

We wish to prove that

Parallel chords intercept equal arcs of the circumfer- ence.

Let the chords A B, C D, be parallel; then will the intercepted arcs A C and B D be equal.

For draw the straight line B C.

Because the lines A B and C D are parallel, the inte- rior alternate angles A B C, B C D, are equal.

But the angle A B C is measured by half the arc A C;

And the angle B C D is measured by half the arc B D:

Then, because the angles are equal, the half arcs which measure them must be equal, and the whole arcs themselves must be equal.

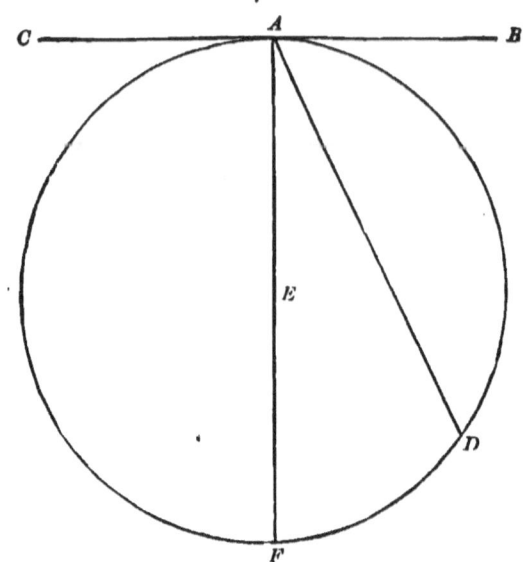

PROPOSITION XXII. THEOREM.

DEMONSTRATION.

We wish to prove that

The angle formed by a tangent and a chord meeting at the point of contact is measured by half the intercepted arc.

Let the tangent C A B and the chord A D meet at the point of contact A; then will the angle B A D be measured by half the intercepted arc A D.

For draw the diameter A E F.

Because A B is a tangent, and A E a radius at the point of contact, the angle B A F is a right angle, and is measured by the semicircle A D F.

Because the angle F A D is at the circumference, it is measured by half the arc D F.

Then the difference between the angles B A F and
D A F, or B A D, must be measured by half the
difference of the arcs A D F and D F, or A D;

That is, the angle B A D is measured by half the
arc A D.

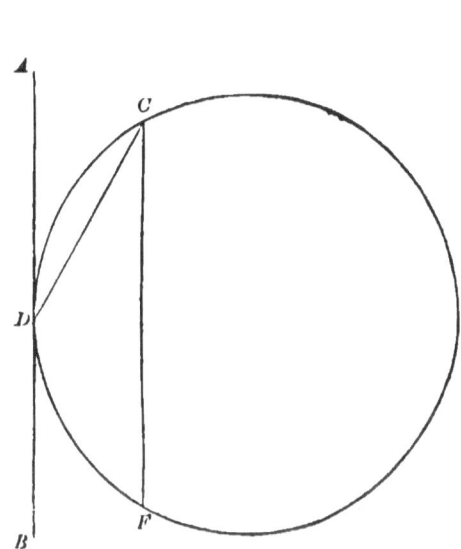

PROPOSITION XXIII. THEOREM.

DEMONSTRATION.

We wish to prove that

*A tangent and chord parallel to it intercept equal arcs
of the circumference.*

Let A B be tangent to the circumference at the
point D, and let C F be a chord parallel to the
tangent; then will the intercepted arcs C D and
D F be equal.

For from the point of contact D, draw the straight line D C.

Because the tangent and chord are parallel, the interior alternate angles A D C and D C F are equal.

But the angle A D C, being formed by the tangent D A and the chord D C, is measured by half the intercepted arc D C;

And the angle D C F, being at the circumference, is measured by half the arc on which it stands, D F:

Then, because the angles are equal, the half arcs which measure them are equal, and the arcs themselves are equal.

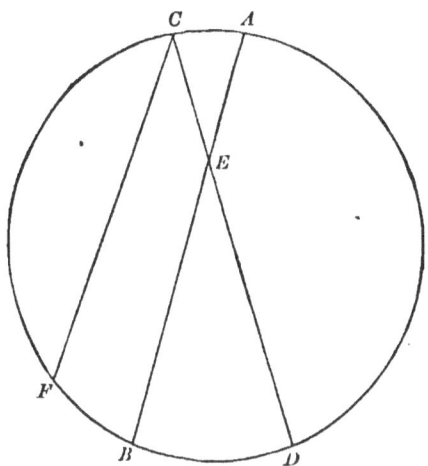

PROPOSITION XXIV. THEOREM.

DEMONSTRATION.

We wish to prove that

The angle formed by the intersection of two chords in a circle is measured by half the sum of the intercepted arcs.

Let the chords A B and C D intersect each other in the point E; then will the angle B E D or A E C be measured by half the sum of the arcs A C, B D.

For from the point C draw C F parallel to A B.

Because the chords A B and C F are parallel, the arcs A C, B F, are equal.

Add each of these equals to B D, and we have B D plus A C equal to B D plus B F; that is, the sum of the arcs B D, A C, is equal to the arc F D.

Because the chords A B, C F, are parallel, the opposite exterior and interior angles D E B, D C F, are equal.

But D C F is an angle at the circumference, and is therefore measured by half the arc F D.

Then the equal angle D E B must be measured by half of the arc F D, or its equal D D, plus A C.

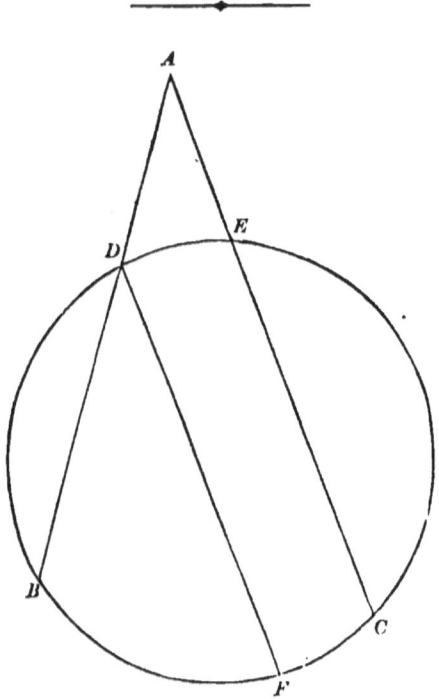

PROPOSITION XXV. THEOREM.

DEMONSTRATION.

We wish to prove that

The angle formed by two secants meeting without a circle is measured by half the difference of the intercepted arcs.

Let the secants A B, A C, intersect the circumference in the points D and E; then will the angle B A C be measured by half the difference between the arcs B C and D E.

For from the point D draw the chord D F parallel to E C.

Because A C and D F are parallel, the opposite exterior and interior angles B D F and B A C are equal.

Because the chords D F, E C, are parallel, the arcs D E and F C are equal.

If from the arc B C we take the arc D E, or its equal F C, we shall have left the arc B F;

But the angle B D F, being at the circumference, is measured by half the arc B F:

Then the equal of B D F, or B A C, must be measured by half the arc B F, or half the difference between the intercepted arcs B C and D E.

APPENDIX.

NOTE A. — To those teachers who think that the line should be derived from a surface, and the surface from a solid, the author would say, that, according to his experience, children apprehend the ideas conveyed by the terms *line* and *surface* as readily as they do any ideas whatever; and that, therefore, there seems to be no necessity for extraordinary care in this case to avoid giving wrong impressions.·

Still, if it be considered desirable in this manner to derive lines and surfaces, it will be apparent that all that can be done in the matter is to give such instruction only by way of a preliminary lesson.

NOTE B. — Crooked and curved lines are here treated of before straight lines, because the first two are defined by means of an affirmative property, — they *do* change direction; while the last is defined by means of the absence of one, — they do *not* change direction. It is easier for a child to comprehend what *is* than what is *not*.

NOTE C. — If the pupils are old enough, they may be shown that vertical lines cannot be parallel, but only seem so on account of their shortness and nearness to each other.

NOTE D. — This definition may be considered objectionable because *rhomboid* means like a rhomb. That the more general figure, the rhomboid, has been named from the more restricted one, the rhomb, is unfortunate, because it interferes with the symmetry of the nomenclature. The rhomb possesses all the properties of the rhomboid, and should, therefore, when these are considered, be called by the same name; its additional property entitles it to a name which should comprehend the other names. If the rectangle had been called a squaroid, the difficulty would have been repeated.

NOTE E. — If teachers consider it desirable, they may require the class to prove, by way of corollary, such propositions as assert the parallelism of the lines when the interior alternate angles are equal, when the opposite exterior and interior angles are equal, &c., in continuation of what has already been done.